Edexcel Topic Tutor
GCSE Mathematics
Foundation Tier

Student Book and CD-ROM

edexcel
advancing learning, changing lives

D1827409

CONTENTS

A PEARSON COMPANY

Published by:
Edexcel Limited
One90 High Holborn
London WC1V 7BH

www.edexcel.org.uk

Distributed by:
Pearson Education Limited
Edinburgh Gate
Harlow
Essex CM20 2JE

First published 2006
ISBN 978-1-84690-007-5

Cover design by Youngs Design Ltd
Questions and answers typeset by Gaelle Anderson
Topic Tutor model solutions produced by Live Learning Ltd
Series directed by Mark Jordan

With thanks to the series production team:
John Barrett, Kevin Tanner, Richard Taylor
Clare Berryman, Sacha Harmsworth, Michelle Hessels, Helen Pritt, Mark Ralph, Joanna Shock

Printed and bound in Great Britain by Bell and Bain Limited

CD-ROM reproduction in Germany by Optimal

The Publisher's policy is to use paper manufactured from sustainable forests.

The Publisher wishes to draw attention to the Single User Licence Agreement situated at the back of the book. Please read this agreement carefully before installing and using the CD-ROM.

Every effort has been made to ensure that the structure and level of question papers matches the current specification requirements and that solutions and mark schemes are accurate. However, the publisher can accept no responsibility whatsoever for accuracy of any solutions or answers to these questions. Any such solutions or answers may not necessarily constitute all possible solutions.

Introduction

Welcome to Edexcel Topic Tutor, a resource to support learning and revision for the new 2-Tier GCSE Mathematics course.

This book contains a series of topic tests using real past exam questions, which can be tackled in class, as homework or as part of your revision. Each exercise begins with a list of the skills you are practising. Short answers at the back of the book allow you to quickly spot problem areas. Where you need extra help, you can use the accompanying CD-ROM to view animated model answers with commentary from examiners.

Edexcel Topic Tutor can be used to supplement most schemes of work for GCSE Mathematics. Next to each exercise title in this book is a reference to the relevant chapters in the Edexcel GCSE Mathematics Foundation Tier Student Book (Linear course textbook ISBN 1-903-13390-4 and Modular course textbook ISBN 1-903-13398-X).

Getting started

Insert the disk into your CD-ROM drive. Edexcel Topic Tutor should auto-run on your PC. If it does not auto run, click Start, My Computer, and right-click on the CD-ROM icon and choose 'Explore'. Double click on the Tutor.exe application and follow the onscreen instructions.

If you are a Mac user you will need to double-click the CD icon on the desktop and double-click on the Topic Tutor application.

Animated model solutions are provided for all exercises. To see and hear a worked solution to a question you will need to:

1.   Click on the exercise name from the menu of exercises (a numbered list of answers will appear)

2.   Click the relevant answer number/part

This will open the Live Learning Player. You can click through the many steps of a solution or replay it using the control buttons at the bottom of the Live Learning Player.

You can also view brief answers for any question by clicking on the 'Answers' link on the main screen. You will need Adobe® Acrobat® Reader® installed on your machine or network in order to view the answer PDFs. If you do not have Adobe® Acrobat® Reader®, follow the installation instructions on the Help screen.

Taking the Topic Tests

Each exercise includes space in which to write your final answer and some have space in which to show your working.   An icon next to each question indicates whether or not the exercise is taken from a calculator or non-calculator examination and the number of marks available for a question are shown beside each question.

You can check the answers to each question in the answer section at the end of each unit.

Hardware Requirements

- Operating system: Windows 95(OS R2), 98, ME, 2000, NT or XP
- Pentium 400 (IBM Compatible PC) or equivalent PC
- 128 MB RAM or higher
- 16 bit graphics card
- CD-ROM drive (minimum 16 speed recommended)
- SVGA colour monitor and 1024/768 resolution
- Sound card
- At least 100 MB free hard disk space

Mac System requirements

- Operating system: X 10.1.5 or higher
- 500MHz G4 processor
- 256MB of RAM or higher
- 450MB of free hard disk space
- 16 speed CD ROM drive
- 16 bit colour monitor set at 1024/768 resolution

Technical support and service is available from the Ask Edexcel service at www.edexcel.org.uk/ask

Formulae

Area of trapezium = $\frac{1}{2}(a+b)h$

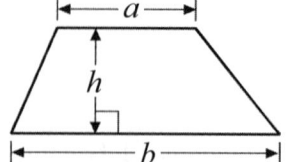

Volume of prism = area of cross section × length

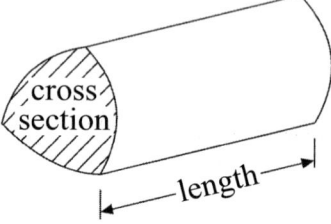

A02
Number
and algebra

Contents

1 Numbers and the number system, solving numerical problems, calculations

1A - Number   LIN 1 MOD 6

- ❏ Read and write whole numbers in digits and in words
- ❏ Order whole numbers including negative numbers in context
- ❏ Understand the value of each digit in a number
- ❏ Round numbers to nearest 10, 100 or 1000
- ❏ Use a timetable or mileage chart
- ❏ Read scales from dials and meters

- ❏ Recognise odd, even and square numbers, factors and multiples
- ❏ Select a multiple of a given number from a list of numbers
- ❏ Select a square number from a list of numbers
- ❏ Select a prime numbers from a list
- ❏ Find the HCF of two numbers

1. (a) Write the number **seventeen thousand, two hundred and fifty-two** in figures.

....
(1)

(b) Write down the value of the 4 in the number 274 863

....
(1)

2. Write $7\frac{1}{2}$ million in figures.

....
(1)

3. (a) Write down the number 1540 in words.

....
(1)

(b) Write down the value of the 7 in the number 9704

....
(1)

4.

Natasha had one pound sixty pence.

Her friend, Kelly, had two pounds five pence.

Write down, in figures, how much money
Kelly and Natasha each had.

one pound sixty pence

two pounds five pence

Natasha   £ ....

Kelly   £ ....

(2)

5.

The table shows the lengths, in kilometres, of 5 rivers.

River	Length in kilometres
Volga	3700
Missouri	3726
Rio Grande	3034
Mississippi	3780
Yukon	3185

Write these lengths in order of size.
Start with the shortest length.

.... km   .... km   .... km   .... km   .... km

(2)

A02 - Number and algebra

6. Write these numbers in order of size.
Start with the smallest number.

(a) 76,   103,   13,   130,   67

....

(1)

(b) −3,   5,   0,   −7,   −1

....

(1)

7. Sally wrote down the temperature at different times on 1st January.

Time	Temperature
midnight	− 6 °C
4 am	− 10 °C
8 am	− 4 °C
noon	7 °C
3 pm	6 °C
7 pm	− 2 °C

Write down

(i) the **highest** temperature,

........................°C

(ii) the **lowest** temperature.

........................°C

(2)

8. Write these numbers in order of size.

Start with the smallest number.

(i) 5, −6, −10, 2, −4

....

(ii) $\frac{1}{2}, \frac{2}{3}, \frac{2}{5}, \frac{3}{4}$

....

(3)

9. There are three cards with numbers on.
The cards are placed to make the number 419

| 4 | 1 | 9 |

(i) Write the numbers 4, 1, 9 on the cards below to give the **highest** possible number.

☐ ☐ ☐

(1)

(ii) Write the numbers 4, 1, 9 on the cards below to give the **lowest** possible number.

☐ ☐ ☐

(2)

A02 - Number and algebra

10.

(i)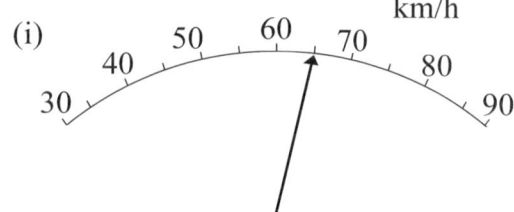

(ii)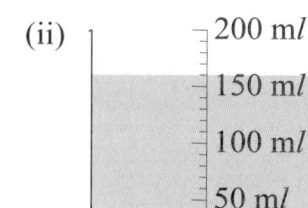

(iii)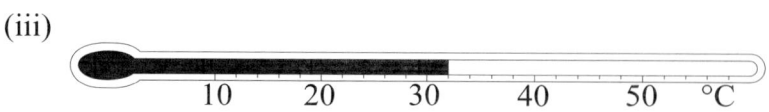

Write down the reading on each of these scales.

(i) .... km/h

(ii) .... m*l*

(iii) .... °C

(3)

11.

Here is a thermometer.
The thermometer shows Nazia's temperature, in °C.

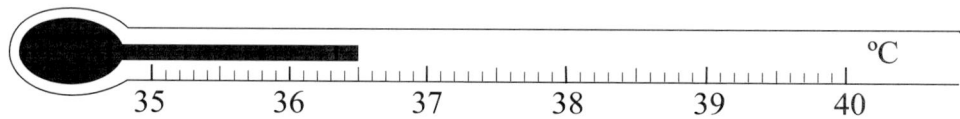

(a) Write down Nazia's temperature.

.... °C
(1)

Nazia becomes ill.
Her temperature reaches 38.4°C.

(b) Show a temperature of 38.4°C on the thermometer.

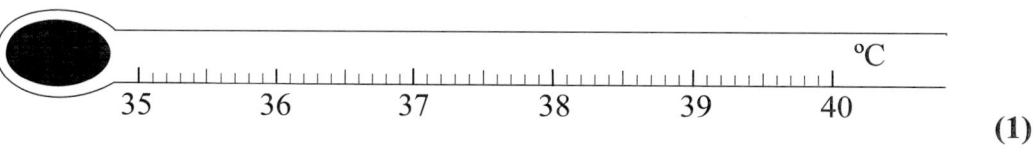

(1)

A02 - Number and algebra

12.

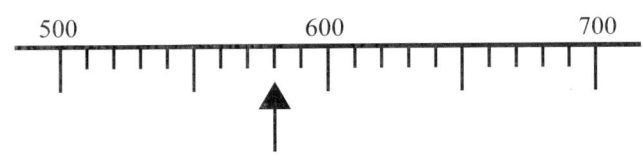

(a) Write down the number marked with an arrow.

.... (1)

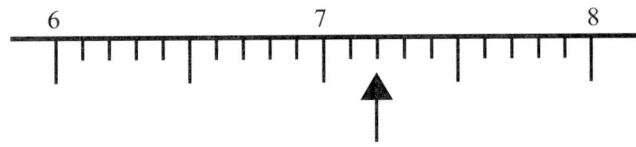

(b) Write down the number marked with an arrow.

.... (1)

(c) Find the number 48 on the number line.
Mark it with an arrow (↑).

(1)

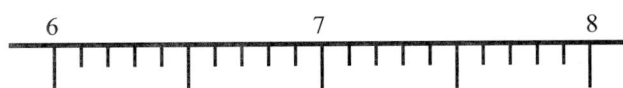

(d) Find the number 6.7 on the number line.
Mark it with an arrow (↑).

(1)

13. Here are two readings from a gas meter.

| 0 | 1 | 9 | 6 | 2 |

January

| 0 | 2 | 1 | 5 | 9 |

April

The difference in the meter readings gives the number of units of gas used.

(a) Work out the number of units of gas used.

....
(2)

The cost of each unit of gas is 21p.

(b) Work out the cost of the gas used between January and April.
Give your answer in pounds (£).

£ ....
(2)

14. Here is a list of 8 numbers.

11   16   18   36   68   69   82   88

Write down **two** numbers from the list with a sum of 87

.... , ...
(1)

15. n is a whole number.

What type of whole number is $2n$?

....

(1)

16. Find the Highest Common Factor (HCF) of 84 and 180

....

(2)

1B - Calculations

❏ Add and subtract negative numbers   ❏ Use long division

❏ Use long multiplication   ❏ Use a calculator efficiently

1. The table shows the temperature on the surface of each of five planets.

Planet	Temperature
Venus	480 °C
Mars	− 60 °C
Jupiter	− 150 °C
Saturn	− 180 °C
Uranus	− 210 °C

(a) Work out the difference in temperature between Mars and Jupiter.

........................°C

(1)

(b) Work out the difference in temperature between Venus and Mars.

........................°C

(1)

(c) Which planet has a temperature 30°C higher than the temperature on Saturn?

....

(1)

The temperature on Pluto is 20°C lower than the temperature on Uranus.

(d) Work out the temperature on Pluto.

........................°C

(1)

2. Work out 286×43

....

(3)

3. The cost of a calculator is £6.79

Work out the cost of 28 of these calculators.

£ ....

(3)

4. Enzo makes pizzas.

One day he makes 36 pizzas.
He charges £2.45 for each pizza.

(a) Work out the total amount he charges for 36 pizzas.

£.........................
(3)

Mario delivers pizzas.
He is paid 65p for each pizza he delivers.
One day he was paid £27.30 for delivering pizzas.

(b) How many pizzas did Mario deliver?

.... pizzas
(3)

5. Simon drives 28 miles every day.

There were 365 days in 2005.

How many miles did Simon drive in 2005?

.... miles
(3)

6.

> ┌─────────────────────┐
> │ **Canal boat for hire** │
> │ **£1785.00** │
> │ **for 14 days** │
> └─────────────────────┘

What is the cost **per day** of hiring the canal boat?

£......................

(3)

7. Use your calculator to work out the value of $\dfrac{8.95 + \sqrt{7.84}}{2.03 \times 1.49}$

(a) Write down all the figures on your calculator display.

....

(2)

(b) Write down your answer to part (a) correct to 3 significant figures.

....

(1)

8. (a) Work out the value of $3.8^2 - \sqrt{75}$
Write down all the figures on your calculator display.

....

(2)

(b) Write your answer to part (a) correct to 1 significant figure.

....

(1)

9. Use your calculator to work out

$$5.2 + \sqrt{7.84}$$

....

(2)

1C - Powers and indices

LIN 1, 20 MOD 6, 19

- ❏ Work out simple squares, cubes and roots of numbers
- ❏ Understand the meaning of the terms square, cube, square root
- ❏ Understand the rules of indices when applied to numbers
- ❏ Find the reciprocal of a number
- ❏ Solve problems with powers

1.

Work out the cube of 6

....

(1)

2.

$$2 \qquad 3 \qquad 4 \qquad 5 \qquad 6 \qquad 7 \qquad 8$$

From the list of numbers, write down

(i) the square number,

....

(ii) the cube number,

....

(iii) the square root of 9.

....

(3)

3.

(a) Write as a power of 5

    (i) $5^4 \times 5^2$

....

    (ii) $5^9 \div 5^6$

....

    (2)

(b) $2^x \times 2^y = 2^{10}$

    and

$2^x \div 2^y = 2^4$

Work out the value of x and the value of y.

$x = $ ....

$y = $ ....

    (3)

A02 - Number and algebra

4. Write as a power of 7

(i) $7^5 \times 7^3$

....

(ii) $7^{10} \div 7^4$

....

(2)

5. Write as a power of 7

$$\dfrac{7^5 \times 7^3}{7^{10} \div 7^4}$$

....

(3)

6. Write down the reciprocal of 4

....

(1)

7. The number 40 can be written as $2^m \times n$, where m and n are prime numbers.

Find the value of m and the value of n.

$m = $ ....

$n = $ ....

(2)

❏ Order a list of fractions
❏ Convert from a fraction to a decimal
❏ Convert from a fraction to a recurring decimal
❏ Convert from a fraction to a percentage

A02 - Number and algebra

1.

   (a) Work out $\frac{4}{5}$ of 30

.... **(2)**

   (b) Write $\frac{4}{5}$ as a decimal.

.... **(2)**

2.

Write these numbers in order of size.

Start with the smallest number.

$$\frac{1}{2}, \ \frac{2}{3}, \ \frac{2}{5}, \ \frac{3}{4}$$

.... **(3)**

3.

$$\frac{3}{5} \qquad \frac{3}{7} \qquad \frac{3}{8} \qquad \frac{3}{10} \qquad \frac{3}{11}$$

Bronwyn converted each of these fractions to decimals.
Some of these fractions gave a recurring decimal.

Put a ring around each of these fractions.

(2)

4.

$$\frac{1}{3} \qquad \frac{2}{5} \qquad \frac{5}{8} \qquad \frac{6}{10} \qquad \frac{7}{12} \qquad \frac{9}{15}$$

Maria correctly converted each of these fractions to decimals.

Put a ring around each fraction which gave a recurring decimal.

(2)

5.

Change $\frac{3}{11}$ to a decimal.

....

(1)

6.

Write $\frac{1}{4}$ as a percentage.

.....................%

(1)

1E - Calculations with fractions

LIN 13 MOD 26

❑ Work out the fraction of quantities

❑ Simplify a fraction by cancelling

❑ Express a given number as a fraction of another

❑ Calculate with fractions

❑ Add and subtract fractions with no mixed numbers

❑ Add and subtract fractions with mixed numbers

❑ Multiply and divide fractions: no mixed numbers

A02 - Number and algebra

1. Alison travelled a total of 145 miles.

She travelled $\frac{2}{5}$ of this distance in the morning.

How many miles did she travel during the rest of the day?

.... miles

(3)

2.

Tigers Club

Admission:
£2.40
Special offer
20% off

Cheetahs Club

Admission:
£2.70
Special offer
$\frac{1}{3}$ off

It normally costs £2.40 to get into the Tigers Club but there is 20% off the price.

It normally costs £2.70 to get into the Cheetahs Club but there is $\frac{1}{3}$ off the price.

Which club is cheaper?
You **must** show all your working with your answer.

....

(4)

3.

(a) Work out $\frac{1}{3}$ of 21

....
(1)

(b) Work out $\frac{3}{5}$ of 35

....
(2)

4.

(a) Work out $\frac{2}{5} + \frac{1}{10}$

....
(2)

(b) Work out $\frac{2}{3} \times \frac{1}{4}$

Write your answer as a fraction in its simplest form.

....
(2)

5.

Some students each chose one PE activity.

$\frac{1}{5}$ of the students chose swimming.

$\frac{3}{8}$ of the students chose tennis.

All the rest of these students chose cricket.

What fraction of the students chose cricket?

....

(3)

6.

Work out $4\frac{1}{2} + 1\frac{2}{5}$

....

(3)

7.

    (a) Work out the value of $\frac{2}{3} \times \frac{3}{4}$

        Give your answer as a fraction in its simplest form.

        ....

        (2)

    (b) Work out the value of $1\frac{2}{3} + 2\frac{3}{4}$

        Give your answer as a fraction in its simplest form.

        ....

        (3)

8.

    Work out $60 \times \frac{2}{3}$

    ....

    (2)

1F - Decimals and percentages

LIN 7, 17 MOD 10, 18

❏ Order a list of decimals

❏ Find a percentage of a quantity

❏ Convert from a decimal to a fraction or a percentage

❏ Convert from a percentage to a decimal or a fraction

❏ Understand that a whole is equivalent to 100%

❏ Order a short list of common fractions, decimals and percentages

❏ Order a long list of fractions, decimals and percentages

❏ Calculate VAT

❏ Increase and decrease quantities by a given percentage

❏ Add on VAT

---

1. Write these numbers in order of size.
Start with the smallest number.

0.72,   0.7,   0.072,   0.07,   0.702

....
(1)

---

2. Work out 70% of £340

£...................................
(2)

---

3. Mrs Brown gets a 25% reduction if she spends £120 or more.

Work out 25% of £120

£......................
(2)

4.   (a)   Work out 50% of £640

£ ....
(2)

(b)   Work out 10% of £56

£ ....
(2)

5.   Linda gets 24 out of 40 in a science test.

Write 24 out of 40 as a percentage.

.... %
(2)

6.   (a)   Write 0.85 as a percentage.

....%
(1)

(b)   Write $\frac{1}{10}$ as a percentage.

....%
(1)

(c)   Write 60% as a decimal.

....
(1)

7.

> **Cat facts**
>
> - 40% of people named cats as their favourite pet.
>
> - 98% of women said they would rather go out with someone who liked cats.
>
> - About $7\frac{1}{2}$ million families have a cat.
>
> - $\frac{1}{4}$ of cat owners keep a cat because cats are easy to look after.

(a) Write 40% as a fraction.
Give your fraction in its simplest form.

....
(2)

(b) Write 98% as a decimal.

....
(1)

8.

Write as a fraction

63%

....
(1)

9.

Write these numbers in order of size.
Start with the smallest number.

22% $\frac{1}{5}$ 0.3 $\frac{2}{7}$

....
(3)

10. Alistair sells books.
He sells each book for £7.60 plus VAT at $17\frac{1}{2}$ %.

He sells 1650 books.

Work out how much money Alistair receives.

£......................

(4)

11. The price of a DVD player was £120
In a sale, the price is reduced by 35%.

Work out the sale price of the DVD player.

£...............................

(3)

12. A compact disc player costs £50 plus 17½% VAT.

Calculate the total cost of the compact disc player.

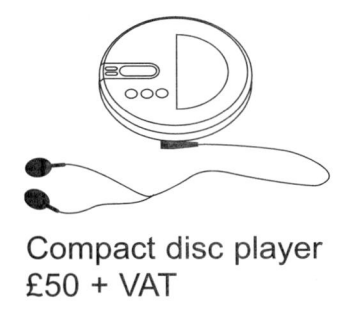

Compact disc player
£50 + VAT

£....................................

(3)

1G - Money

❑ Calculate shopping bills and work out the amount of change

❑ Work out how many items can be bought for a given amount of money

❑ Calculate a gas or electricity bill given meter readings and charges

❑ Use decimals to work out total cost given the cost of one part

❑ Use exchange rates to work out equivalent amounts in different currencies

1. Michael buys 3 files.

The total cost of these 3 files is £5.40

Work out the total cost of 7 of these files.

£......................

(3)

2. Rizwan buys

6 stamps at 25p each
2 packs of postcards at 89p per pack
1 pack of labels at £1.09

He pays with a £10 note.

Work out how much change Rizwan should get.

£......................

(3)

3.   The table below shows the cost of each of three calculators.

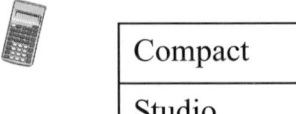

Compact	£2.30
Studio	£2.15
Basic	£2.80

Barbara buys one Studio calculator and one Compact calculator.
She pays with a £10 note.

(a) How much change should she get?

£.........................
(4)

Mrs Brown wants to buy some Basic calculators.
She has £60 to spend.

(b) Work out the greatest number of Basic calculators she can buy.

....
(2)

4.   The cost of a compact disc holder is 25p.
John has £15 to spend.

What is the greatest number of compact disc holders that John can buy for £15?

....
(3)

5. Alison travels by car to her meetings.
Alison's company pays her 32p for each mile she travels.

One day Alison writes down the distance readings from her car.

Start of the day: 2430 miles
End of the day: 2658 miles

Work out how much the company pays Alison for her day's travel.

£......................
(4)

6. Here are two readings from a gas meter.

| 0 | 1 | 9 | 6 | 2 |

| 0 | 2 | 1 | 5 | 9 |

January   April

The difference in the meter readings gives the number of units of gas used.

(a) Work out the number of units of gas used.

....
(2)

The cost of each unit of gas is 21p.

(b) Work out the cost of the gas used between January and April.
Give your answer in pounds (£).

£ ....
(2)

7.   The cost of 4 kg of apples is £3.36

The total cost of 3 kg of apples and 2.5 kg of pears is £4.12

Work out the cost of 1 kg of pears.
Give your answer in pence.

..........................p

(3)

8. Sangita is on holiday in Switzerland.
She buys a train ticket.

She can pay either 100 Swiss Francs or 70 Euros.

£1 = 2.10 Swiss Francs
£1 = 1.40 Euros

She pays in Swiss Francs rather than Euros.
Work out how much she saves.
Give your answer in pounds.

£ ....

(4)

A02 - Number and algebra

9. The table can be used to convert between Euros (€) and Pounds (£).

Euros (€)	Pounds (£)
0.10	0.08
0.20	0.16
0.50	0.40
1	0.80
2	1.60
3	2.40
4	3.20

(a) Change €3 to pounds.

£........................

(1)

(b) Change €2.50 to pounds.

£........................

(2)

(c) Change £1 to euros.

€........................

(2)

10. A group of students visited the USA.
A student bought a pair of sunglasses in the USA.
He paid $35.50

In England, an identical pair of sunglasses costs £26.99
The exchange rate was £1 = $1.42

(a) In which country were the sunglasses cheaper?

....

(2)

(b) How much cheaper?

....

(2)

1H - Ratio and proportion   *LIN 24 MOD 33*

- ❏ Work out simple proportion calculations
- ❏ Divide a quantity in the ratio 1 : n or write as a ratio
- ❏ Divide a quantity in the ratio a : b
- ❏ Divide a quantity in the ratio a : b : c
- ❏ Work out the proportional amounts of ingredients in recipes & ratios

1. There are 30 students on a school trip.
12 of the students are girls.
The rest are boys.

(a) Find the ratio of the number of girls to the number of boys.
Give your ratio in its simplest form.

....

(2)

On another school trip, the ratio of the number of teachers to the number of students was 1:7
The **total** number of teachers and students was 160.

(b) Work out the number of teachers on this school trip.

....

(2)

2. The length of a coach is 15 metres.

Jonathan makes a model of the coach.
He uses a scale of 1:24

Work out the length, in centimetres, of the model coach.

.... cm

(2)

3. There are 21 questions in a science test.
Each question is on biology or on chemistry or on physics.

The numbers of questions on biology, chemistry and physics are in the ratios 4 : 2 : 1

(i) What fraction of the questions are on chemistry?

....

(ii) Work out the number of questions that are on biology.

....

(5)

4. Ruth makes poached peaches.

Here is a list of ingredients for making poached peaches for 6 people.

> **Poached Peaches**
>
> Ingredients for 6 people
>
> 12   yellow cling peaches
> 1400 m*l*   water
> 130 g   granulated sugar

Ruth makes poached peaches for 9 people.

Work out the amount of each ingredient needed to make poached peaches for 9 people.

....   yellow cling peaches

.... ml   water

.... g   granulated sugar

(2)

1J - Estimations

❏ Estimate a solution to a calculation by rounding numbers to 1 s.f.

❏ Write a number correct to 1 significant figure

1. Work out an estimate for the value of $\dfrac{5.79 \times 312}{0.523}$

....

(3)

2. Work out an estimate for the value of $\dfrac{637}{3.2 \times 9.8}$

....

(2)

2 Equations, formulae and identities

2A - Introducing algebra

LIN 8 MOD 11

- ❑ Derive an expression
- ❑ Derive expressions (2 or more terms)
- ❑ Simplify terms
- ❑ Simplify by collecting like terms

- ❑ Multiply out brackets
- ❑ Multiply out brackets and simplify
- ❑ Factorise a simple expression with one term
- ❑ Simplify algebra (by addition or subtraction of indices)

1. Andrew, Brenda and Callum each collect football stickers.

Andrew has x stickers.
Brenda has three times as many stickers as Andrew.

(a) Write down an expression for the number of stickers that Brenda has.

....
(1)

Callum has 9 stickers less than Andrew.

(b) Write down an expression for the number of stickers that Callum has.

....
(1)

2. *ABC* is an isosceles triangle.

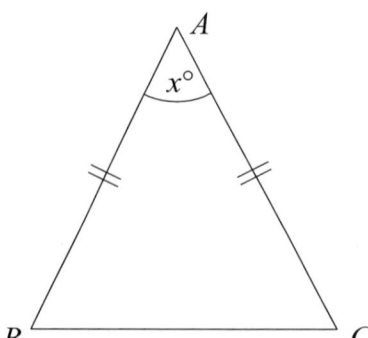

Diagram **NOT**
accurately drawn

$AB = AC$
$AB = 3p + q$
$BC = p + q$

(a) Find an expression, in terms of p and q, for the perimeter of the triangle.
Give your answer in its simplest form.

....

(2)

Angle $A = x°$

(b) Find an expression, in terms of x, for the size of angle B.

....

(2)

A02 - Number and algebra

3. Jo buys 8 cups and 8 mugs.

   A cup costs £x

   A mug costs £$(x + 2)$

   Write down an expression, in terms of x, for the total cost, in pounds, of 8 cups and 8 mugs.

   £.......................

   (2)

4. $a = 4$

   $b = -3$

   Work out the value of $3a + 2b$

   ....

   (2)

5.   (a)   Simplify $5m + 3m - 2m$

....

(1)

     (b)   Simplify $p + 7q + 3p - 2q$

....

(2)

     (c)   Multiply out $3(t - 4)$

....

(1)

     (d)   Simplify $4a \times 3b$

....

(1)

6.   (a)   Simplify $3p + 2q - p + 2q$

....

(2)

     (b)   Simplify $3y^2 - y^2$

....

(1)

     (c)   Simplify $5c + 7d - 2c - 3d$

....

(2)

     (d)   Simplify $4p \times 2q$

....

(1)

7. Simplify

(i) $2c + 3c + 4c$

....

(ii) $f \times g \times 3$

....

(iii) $x^2 + x^2 + x^2$

....

(3)

8. Simplify $3f + 2g - f + 5g$

....

(2)

9.

(a) Simplify

    (i) $3a + 4b - 2a - b$

....

    (ii) $5x^2 + 2x - 3x^2 - x$

....

(4)

(b) Expand the brackets

    (i) $4(2x - 3)$

....

    (ii) $p(q - p^2)$

....

(2)

(c) Expand and simplify $5(3p + 2) - 2(5p - 3)$

....

(2)

A02 - Number and algebra

10.    (a) Simplify $4a + 5b - 3b + a$

....
(2)

(b) Simplify $x^3 + x^3$

....
(1)

11.   (a) Expand $3(t - 4)$

....
(1)

(b) Factorise completely $6q + 12$

....
(2)

12.   Expand $y(y^3 + 2y)$

....
(2)

13.   Expand and simplify $2(3y + 4) + 3(y - 1)$

....
(2)

14.   Simplify $p^8 \div p^2$

....
(1)

| 2B - Expanding and factorising quadratics | LIN 27 MOD 21 |

❏ Expand and simplify $(x + a)(x + b)$ ❏ Factorise simple quadratic expressions

1.   Expand and simplify $(y + 3)(y + 4)$

....
                                                                         (2)

2.   Expand and simplify $(x + 7)(x - 4)$

....
                                                                         (2)

3.   Factorise $x^2 - 3x$

....
                                                                         (2)

4.   Factorise $p^2 + 6p$

....
                                                                         (2)

Factorise completely $6x^2 - 9xy$

....
                                                                         (2)

5. Factorise fully $2p^2 - 4pq$

....

(2)

2C - Formulae

- ❏ Use formulae in words
- ❏ Use simple rules
- ❏ Substitute into formulae/expressions
- ❏ Use negative integers in formulae
- ❏ Substitute into quadratic expressions

- ❏ Devise a rule or a formula in words and letters
- ❏ Derive a formula with at least two unknowns
- ❏ Use inverse rules with formulae
- ❏ Rearrange formulae

1. Michelle makes dresses.
She uses this rule to work out her pay.

> Multiply the number of dresses made by £5 and add £21

Michelle made 6 dresses.

(a) Use the rule to work out her pay.

£....................................

(2)

Andy also makes dresses.
He is paid using the formula

> $P = 4n + 32$

P is his pay in pounds (£).
n is the number of dresses he makes.

Andy made 7 dresses.

(b) Use the formula to work out his pay.

£....................................

(2)

2.   The cost, in pounds, of hiring a car can be worked out using this rule.

> Add 3 to the number of days' hire
>
> Multiply your answer by 10

(a) Work out the cost of hiring a car for 4 days.

£........................
(2)

Bishen hired a car.
The cost was £120

(b) Work out the number of days for which Bishen hired the car.

....
(2)

3.   The heat setting number of a gas oven is called its Gas Mark.
This rule may be used to change a Gas Mark to a temperature in °C.

> Gas Mark ➤ × 14 ➤ + 121 ➤ Temperature in °C

Use the rule to change Gas Mark 7 to a temperature in °C.

........................°C
(2)

4.

$T = 5p + 3q$

Work out the value of T when $p = 2$ and $q = 4$

$T = $ ....

(2)

---

5. John used this formula to work out his overtime pay.

> overtime pay = overtime rate × number of hours overtime worked

John's overtime rate was £7.20 per hour.
He worked 8 hours overtime.

(a) Work out his overtime pay.

£ ....

(2)

John used this formula to work out his total pay.

> total pay = basic pay + overtime pay

John's basic pay was £234

(b) Work out his total pay.

£ ....

(1)

6.   Expand $3(4x - 1)$

....

(1)

7. $v = 15 - 10t$

     $t = 4$

Work out the value of v.

$v = $ ....

(2)

8.   (a)   Work out the value of $2a + ay$ when $a = 5$ and $y = -3$

....

(2)

     (b)   Work out the value of $5t^2 - 7$ when $t = 4$

....

(3)

9. $p = 2$

Work out the value of $5p^3$

....

(2)

10. The cost, in pounds, of hiring a car can be worked out using this rule.

| Add 3 to the number of days' hire |
| Multiply your answer by 10 |

The cost of hiring a car for n days is C pounds.

Write down a formula for C in terms of n.

(3)

11.

Cinema Ticket Prices	
Adults	£4
Child	£3

An adult ticket costs £4.
A child ticket costs £3.

(a) Write down a formula for the total cost, £T, for n adult tickets and c child tickets.

....

           (3)

Hina spends £47 on cinema tickets.
She buys 8 adult tickets.

(b) Work out how many child tickets she buys.

....

           (3)

12.

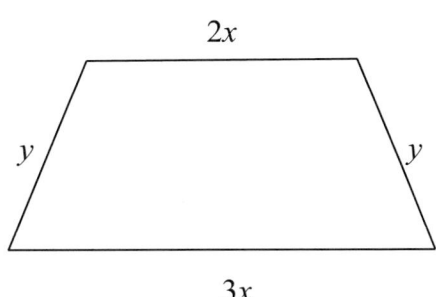

Diagram **NOT** accurately drawn

The diagram shows a trapezium.
All the lengths are in centimetres.
The perimeter of the trapezium is P cm.

Find a formula, in terms of x and y, for P.
Give your answer in its simplest form.

$P =$ ....

(2)

---

13. Rearrange $a(q - c) = d$ to make q the subject.

$q =$ ....

(3)

This publication may be reproduced only in accordance with Edexcel Limited copyright policy. © 2006 Edexcel limited

14. Make t the subject of the formula $\quad v = u + 5t$

$t = $ ....

(2)

15. Make h the subject of the formula

$f = g + 3h$

$h = $ ....

(2)

2D - Equations and inequalities   *LIN 19 MOD 29*

❏ Derive equation from diagrams

❏ Derive and solve equations

❏ Derive and solve equations from a diagram

❏ Solve equations

❏ Solve equations by trial & improvement methods

❏ Solve inequalities

---

1. The width of a rectangle is x centimetres.
The length of the rectangle is $(x + 4)$ centimetres.

$x + 4$

x

(a) Find an expression, in terms of x, for the perimeter of the rectangle.
Give your expression in its simplest form.

....

(2)

The perimeter of the rectangle is 54 centimetres.

(b) Work out the length of the rectangle.

.... cm

(3)

---

2. Imran thinks of a number.

He multiplies the number by 3.
He then adds 19.

His answer is 61.

What number did Imran first think of?

....

(2)

3.

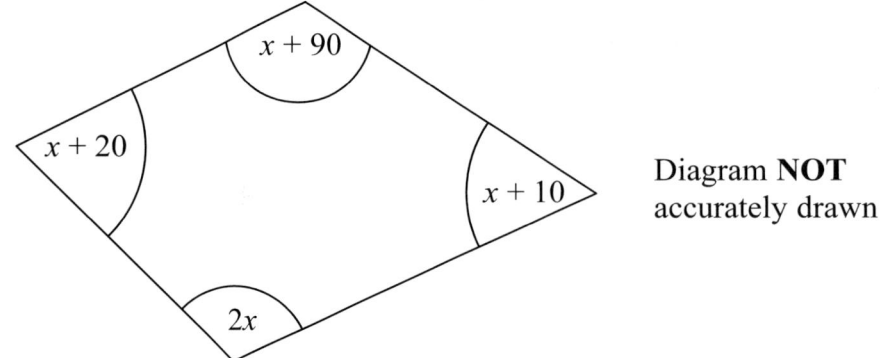

The sizes of the angles, in degrees, of the quadrilateral are

$x + 10$
$2x$
$x + 90$
$x + 20$

(a) Use this information to write down an equation in terms of x.

....

(2)

(b) Use your answer to part (a) to work out the size of the smallest angle of the quadrilateral.

.... °

(3)

4. The diagram below shows a 6-sided shape.
All the corners are right angles.
All measurements are given in centimetres.

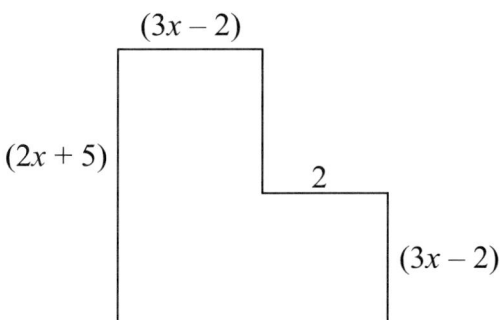

$(3x - 2)$

$(2x + 5)$

2

$(3x - 2)$

Diagram **NOT**
accurately drawn

The area of the shape is 25 cm².

Show that $6x^2 + 17x - 39 = 0$

(3)

5. Solve $x + 8 = 13$

$x = $ ....

(1)

6. (a) Solve $4x - 1 = 7$

$$x = \text{...................}$$
(2)

(b) Solve $5(2y + 3) = 20$

$$y = \text{...................}$$
(3)

7. Solve $\qquad 4(x + 3) = 6$

$$x = \text{...........................}$$
(3)

8. Solve the equation

$$5(x - 3) = 2x - 22$$

$$x = \text{..........................}$$
(3)

9.

(a) Solve $5 - 3x = 2(x + 1)$

$x = $ ....

(3)

(b) $-3 \leqslant y < 3$
y is an integer.

Write down all the possible values of y.

....

(2)

10. Solve $4(y + 3) = 6$

....

(3)

11.   Solve these equations

(a) $x + 5 = 2$

$x =$ ....
(1)

(b) $5p - 3 = 4$

$p =$ ....
(2)

(c) $2q - 4 = 5q + 5$

$q =$ ....
(2)

(d) $5(2r + 7) = 70$

$r =$ ....
(2)

12.   The equation $x^3 + 10x = 21$
has a solution between 1 and 2

Use a trial and improvement method to find this solution.
Give your answer correct to one decimal place.
You must show **ALL** your working.

$x =$ ....
(4)

13.   The equation $x^3 - 4x = 24$

has a solution between 3 and 4.
Use a trial and improvement method to find this solution.
Give your answer correct to 1 decimal place.
You must show **all** your working.

$x =$ ....

(4)

14. $4x + 3y < 12$

x and y are both integers.

Write down two possible pairs of values that satisfy this inequality.

$x =$ .... , $y =$ ....

and $x =$ .... , $y =$ ....

(2)

15.   (i)   Solve the inequality $7x - 3 > 17$

....

(1)

x is a whole number such that $7x - 3 > 17$

(ii)   Write down the smallest value of x.

....

(2)

3 Sequences, functions and graphs

❑ Recognise the pattern in a sequence of diagrams (including finding next in sequence)

❑ Find and use the rule to compute terms in a linear number sequence

❑ Find a term in a linear number sequence

❑ Find missing numbers in a linear series

❑ Find the nth term of a linear expression

1. Here are some patterns made with dots.

Pattern number 1   Pattern number 2   Pattern number 3   Pattern number 4

(a) In the space below, draw Pattern number 5

(1)

(b) How many dots are used in Pattern number 6?

....

(1)

2. (a) The first odd number is 1.

    (i) Find the 3rd odd number.

    ....

    (ii) Find the 12th odd number.

    ....

    (2)

(b) Write down a method you could use to find the 100th odd number.

....

....

(1)

Here are some patterns made with dots.

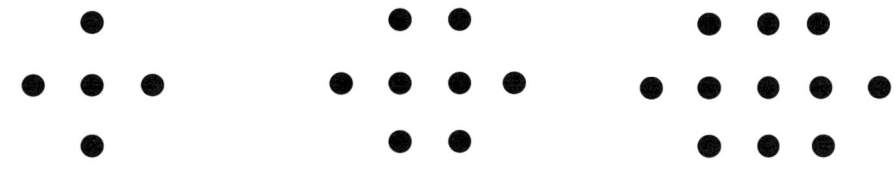

Pattern Number 1   Pattern Number 2   Pattern Number 3

(c) In the space below, complete Pattern Number 4.

(1)

The table shows the number of dots used to make each pattern.

(d) Complete the table.

Pattern Number	1	2	3	4	5
Number of dots	5	8	11		

(2)

3.   Here are some patterns made from matchsticks.

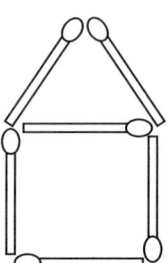   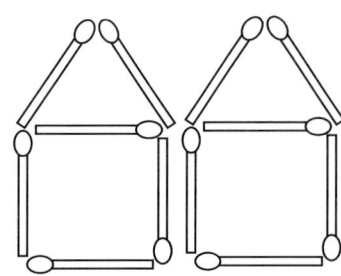   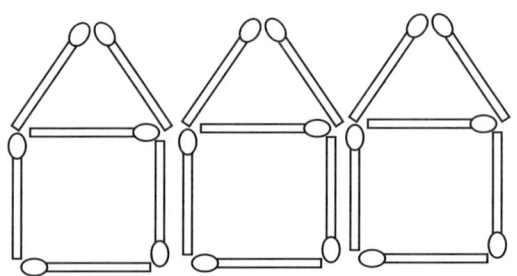

Pattern number 1   Pattern number 2   Pattern number 3

In the space below, draw Pattern number 4

(1)

(b) How many matchsticks are used in Pattern number 10?

....

(1)

(c) Write down a formula for m (the number of matchsticks) in terms of n (the pattern number)

....

(1)

4. Here are the first five terms of a number sequence.

126   122   118   114   110

(a) Write down the next two terms of the number sequence.

.... , ...
(1)

(b) Explain how you found your answer.

....
(1)

The 20th term of the number sequence is 50

(c) Write down the 21st term of the number sequence.

....
(1)

5. Here are the first five terms of a number sequence.

10   16   22   28   34

(a) Write down the next term of the number sequence.

....
(1)

(b) Explain why 861 is **not** a term of the number sequence.

....

....
(1)

6. Here are the first four terms of a number sequence.

$$2 \qquad 7 \qquad 12 \qquad 17$$

(a) Write down the **6th** term of this number sequence.

....

(1)

The nth term of a different number sequence is $4n + 5$

(b) Work out the first three terms of this number sequence.

....

(2)

7. Here are the first five terms of an arithmetic sequence.

$$7 \qquad 11 \qquad 15 \qquad 19 \qquad 23$$

(a) Write down, in terms of n, an expression for the nth term of this sequence.

....

(2)

Pat says that 453 is a term in this sequence.
Pat is wrong.

(b) Explain why.

....

....

(1)

3B - Real life graphs

❏ Plot or read coordinates in 4 quadrants

❏ Interpret a travel graph

❏ Interpret information in a real life graph

❏ Complete travel graphs

❏ Draw and use conversion graphs

1.

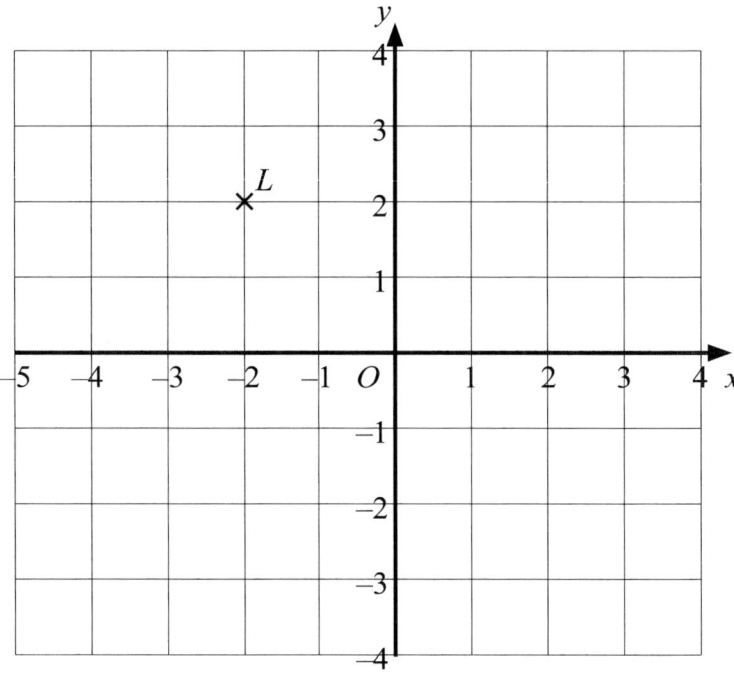

(a) Write down the coordinates of the point *L*.

(............ , ...)
(1)

The coordinates of another point are (–4, –1).

(b) Mark this point on the grid.
Label it *M*.

(1)

2.

Tom travels by car to his meetings.
Tom's company works out the amount it will pay him for the distance he travels.
It uses the graph below.

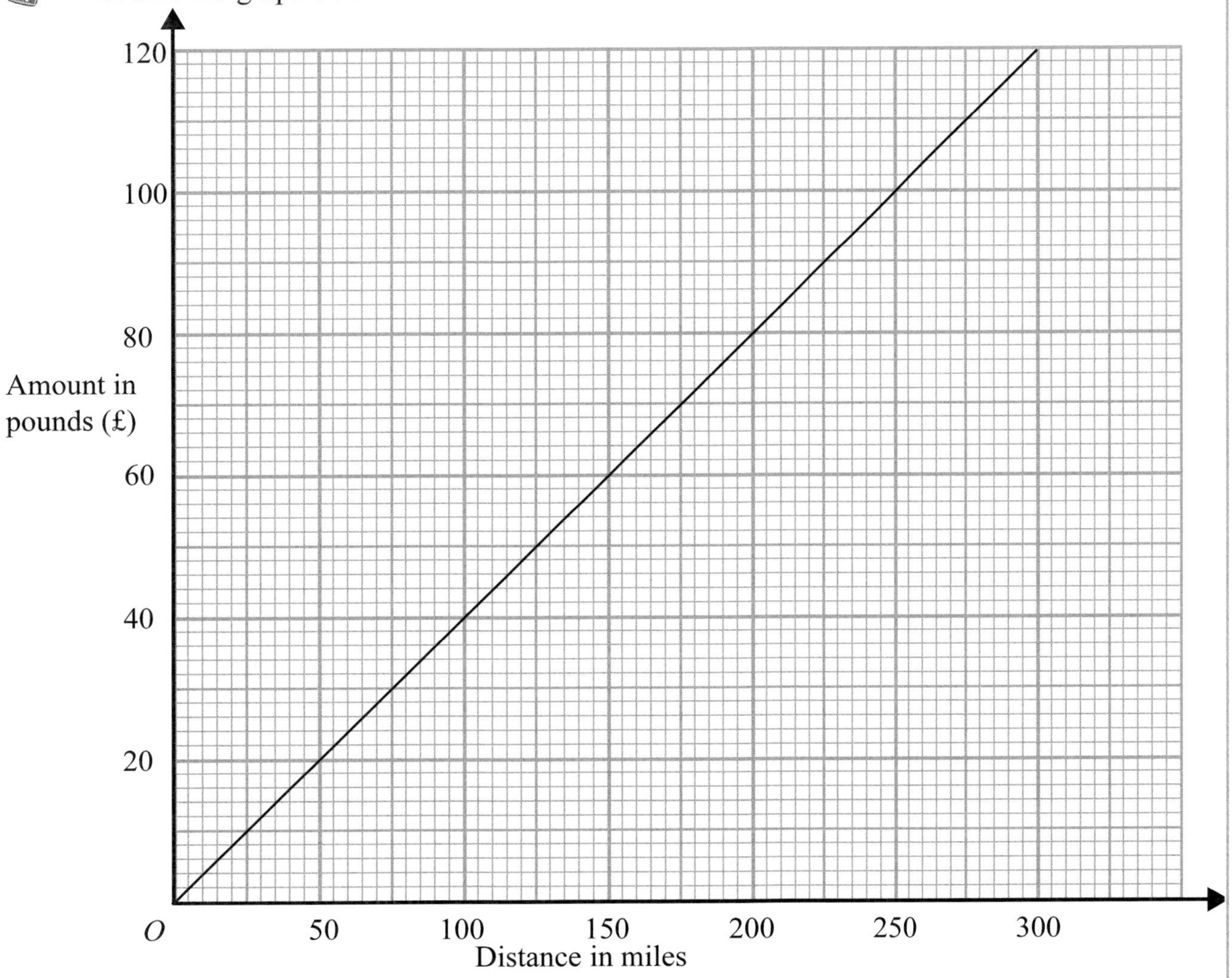

Use the graph to write down

(i) the amount Tom's company pays him when he travels 200 miles,

£.......................

(ii) the distance Tom travels when his company pays him £50.

.................miles

(2)

3. A man left home at 12 noon to go for a cycle ride.
The travel graph represents part of the man's journey.

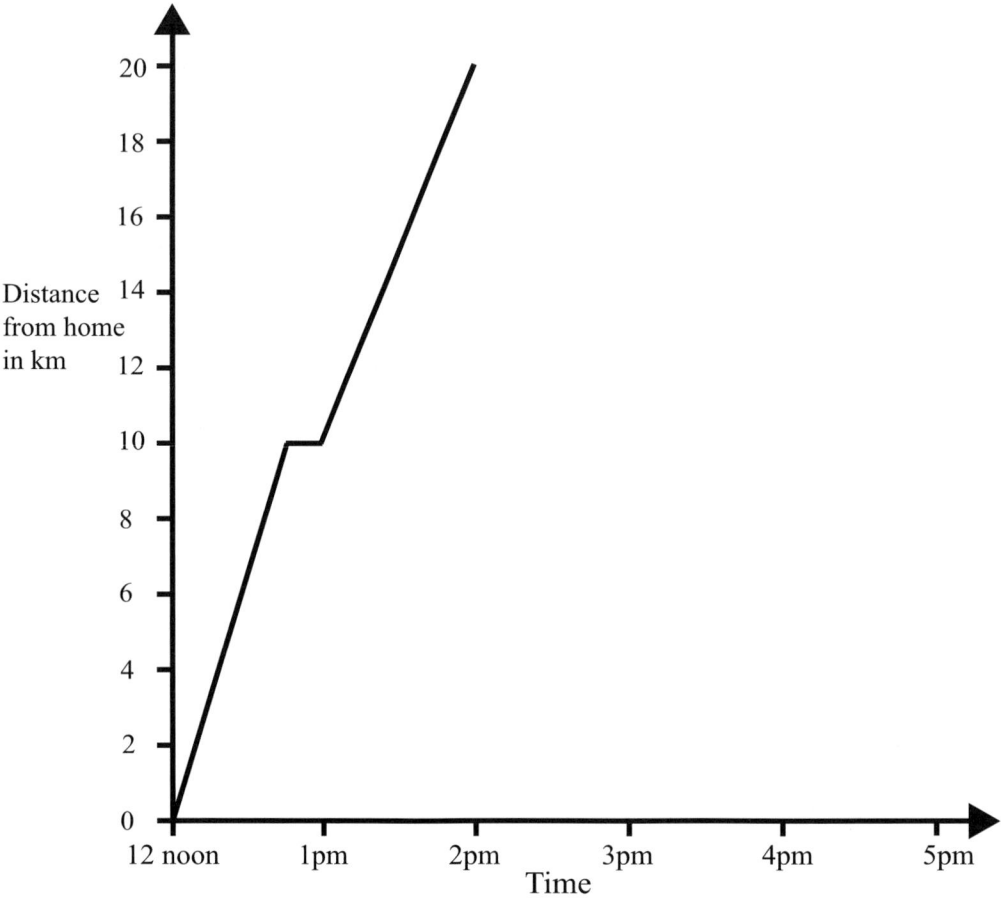

At 12.45pm the man stopped for a rest.

(a) For how many minutes did he rest?

.............minutes

(1)

(b) Find his distance from home at 1.30pm.

.....................km

(1)

The man stopped for another rest at 2pm.
He rested for one hour.
Then he cycled home at a steady speed. It took him 2 hours.

(c) Complete the travel graph.

(2)

4. Here is a conversion graph for changing between kilograms and pounds.

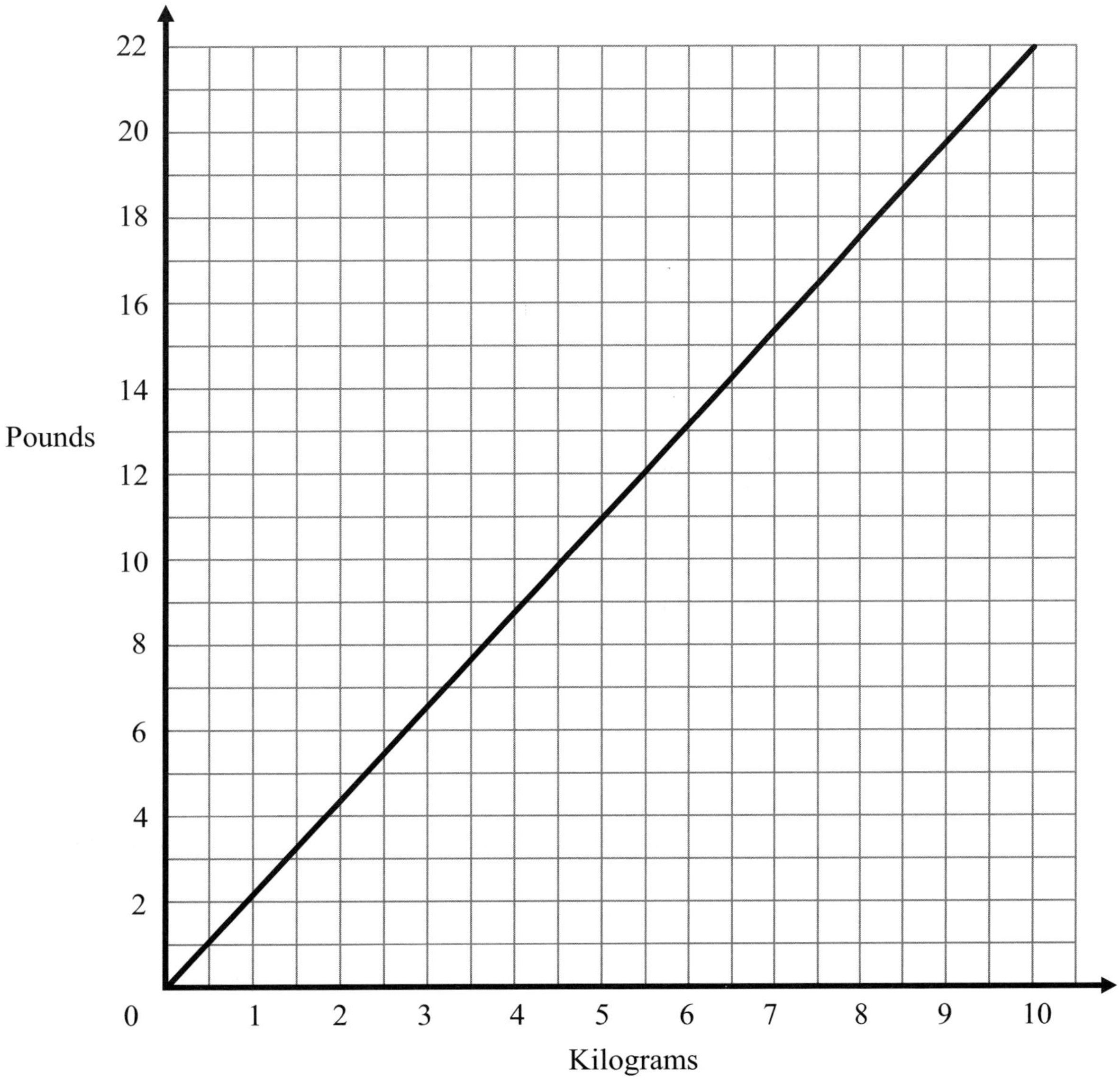

(a) Use the graph to change 22 pounds to kilograms.

.... kg
(1)

(b) Use the graph to change 2.5 kilograms to pounds.

.... pounds
(1)

Fabio weighs 110 pounds.

(c) Change 110 pounds to kilograms.

.... kg
(2)

3C - Graphs on coordinate axes

LIN 15 MOD 16

- ❑ Use a table of values to draw a simple straight line graph
- ❑ Draw the graph of a straight line by finding and plotting points and joining
- ❑ Draw a table of values and the graph of a quadratic function from a table of values
- ❑ Use straight line graphs to solve problems

1. (a) Complete the table of values for $y = 3x + 1$

x	–2	–1	0	1	2	3
y	–5		1			

(2)

(b) On the grid, draw the graph of $y = 3x + 1$

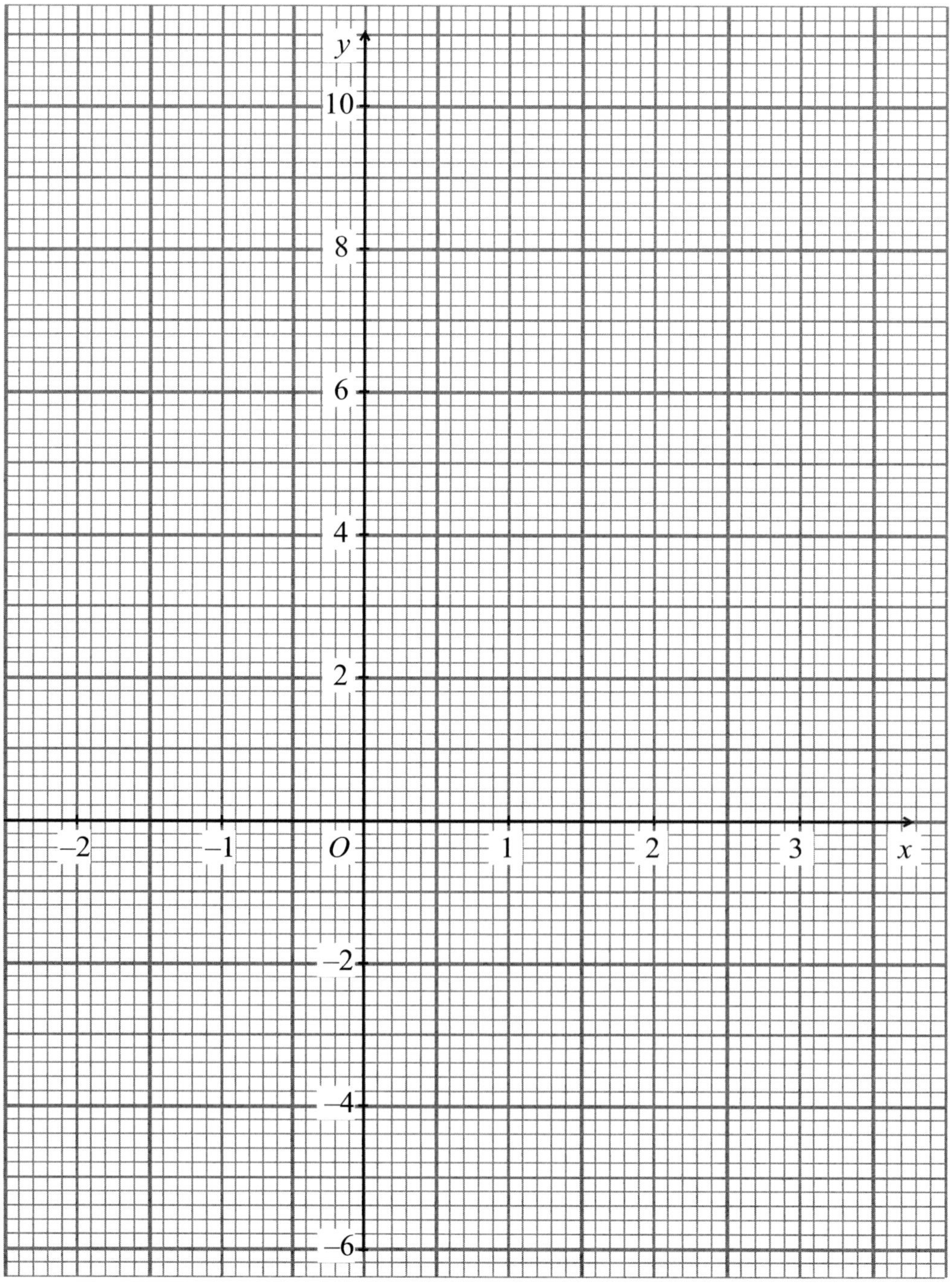

(2)

(c) Use your graph to find

    (i) the value of y when $x = -0.8$

                                            $y = $ ....

    (ii) the value of x when $y = 8.2$

                                            $x = $ ....

(2)

2.   (a) Complete the table of values for $y = 3x + 2$

x	-2	-1	0	1	2
y		-1		5	

(2)

(b) On the grid, draw the graph of $y = 3x + 2$

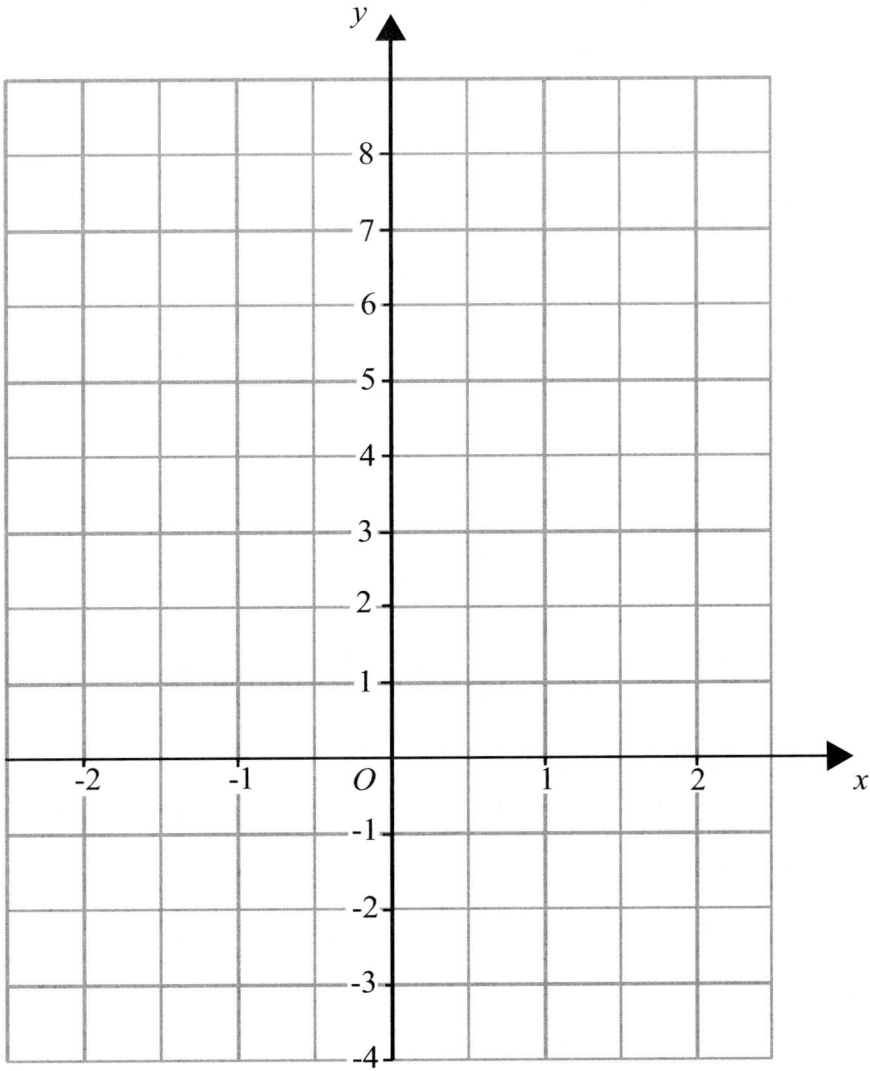

(2)

3.

(a) Complete the table of values for $y = x^2 + x$.

x	-3	-2	-1	0	1	2	3
y	6	2		0		6	

(2)

(b) On the grid, draw the graph of $y = x^2 + x$.

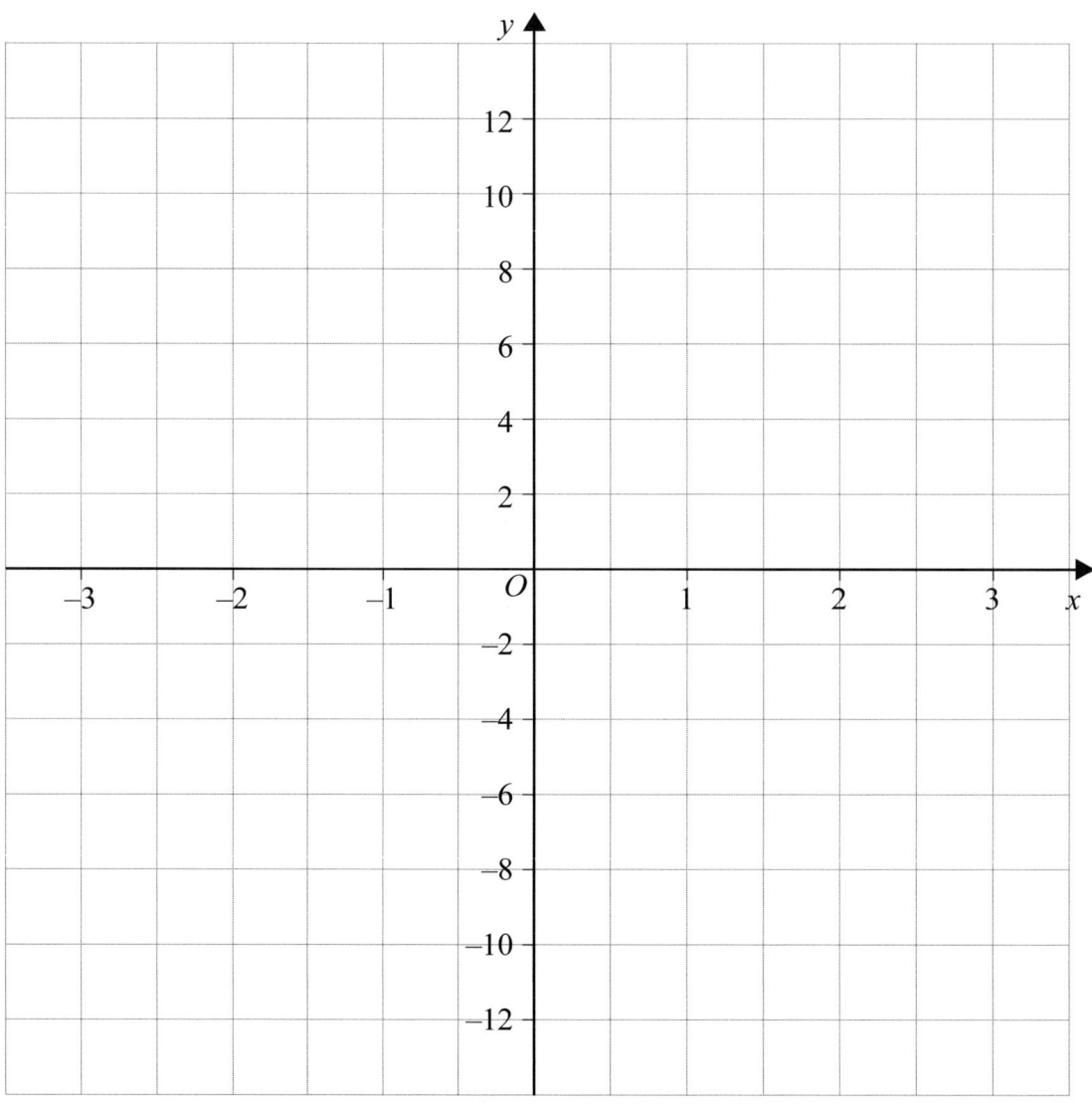

(2)

4.

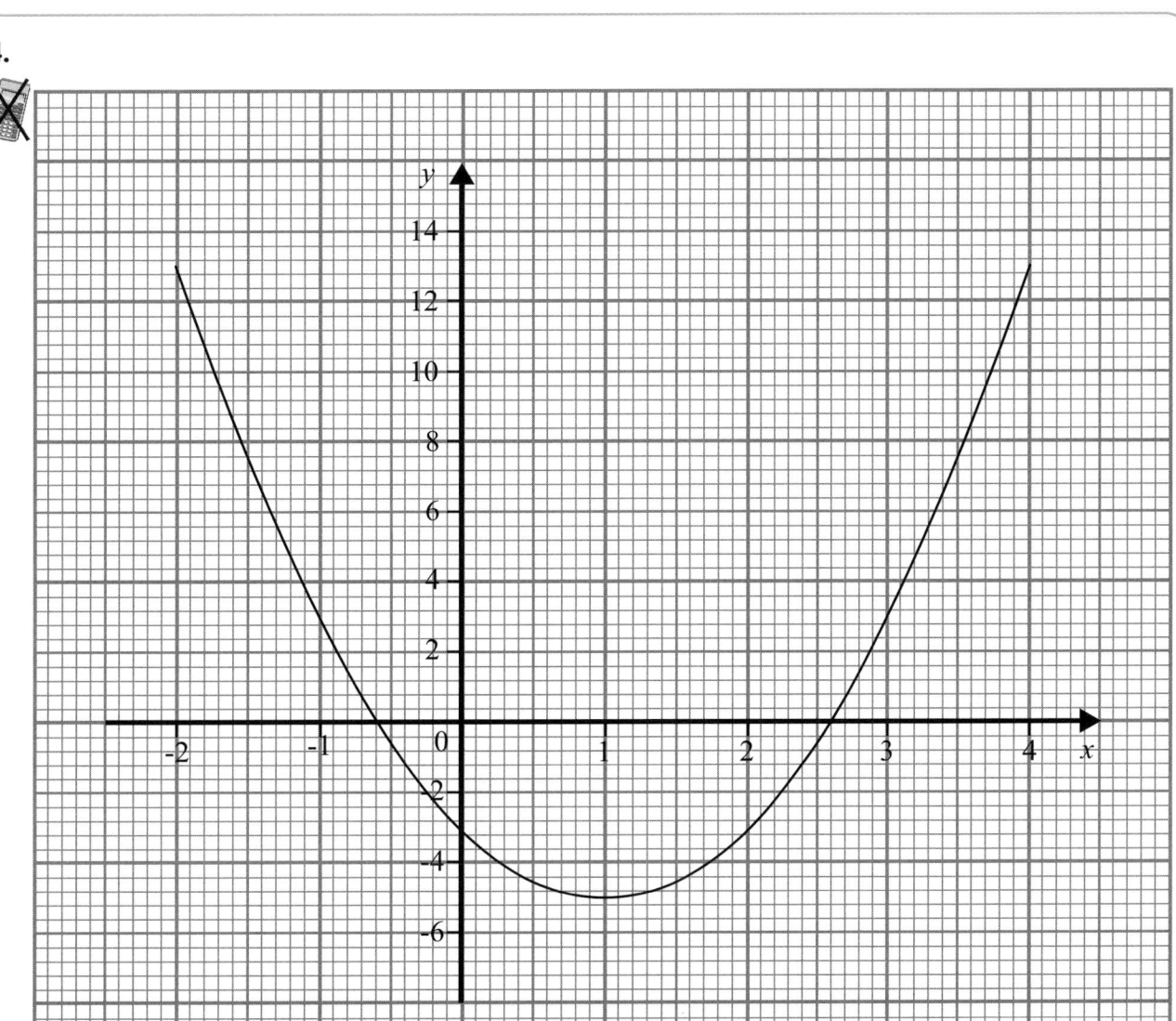

The diagram shows the graph of the equation $y = 2x^2 - 4x - 3$

Use the graph to find the approximate values of x when $2x^2 - 4x - 3 = 0$

$x = $ .... or $x = $ ....

(2)

1A - Numbers

1. (a) 17 252

    (b) thousands, 1000, 4000

2. 7 500 000

3. (a) One thousand five hundred and forty

    (b) 700 or hundreds

4. Natasha: £1.60

    Kelly: £2.05

5. 3034, 3185, 3700, 3726, 3780

6. (a) 13, 67, 76, 103, 130

    (b) −7, −3, −1, 0, 5

7. (i) 7°C

    (ii) −10°C

8. (i) −10, −6, −4, 2, 5

    (ii) $\dfrac{2}{5}, \dfrac{1}{2}, \dfrac{2}{3}, \dfrac{3}{4}$

9. (i) 941

    (ii) 149

10. (i) 65 ± 1

    (ii) 160 ± 2

    (iii) 32 ± 0.5

11. (a) 36.5°C

    (b)
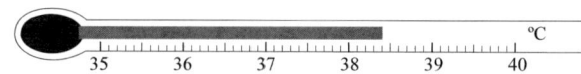

12. (a) 580

    (b) 7.2

    (c)
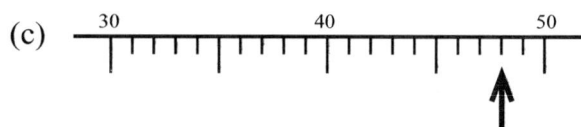

    (d)
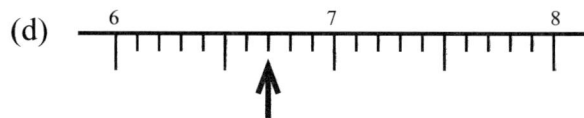

13. (a) 197

    (b) £41.37

14. 18, 69

15. even

16. 12

1B - Calculations

1. (a) 90°C

   (b) 540°C

   (c) Jupiter

   (d) − 230°C

2. 12 298

3. £190.12

4. (a) £88.20

   (b) 42

5. 10 220

6. £127.50

7. (a) 3.884682778

   (b) 3.88

8. (a) 5.77974 5962

   (b) 6

9. 8

1C - Powers and indices

1. 216

2. (i) 4

   (ii) 8

   (iii) 3

3. (a) (i) 5^6

   (ii) 5^3

   (b) $x = 7$

   $y = 3$

4. (i) 7^8
   (ii) 7^6

5. 7^2

6. $\dfrac{1}{4}$ or 0.25

7. $m = 3$

   $n = 5$

1D - Fractions

1. (a) 24

   (b) 0.8(0)

2. $\frac{2}{5}, \frac{1}{2}, \frac{2}{3}, \frac{3}{4}$

3. $\frac{3}{7}, \frac{3}{11}$

4. $\frac{1}{3}, \frac{7}{12}$

5. 0.2727...

6. 25%

1E - Calculations with fractions

1. 87 miles

2. Cheetahs at £1.80
   (Tigers £1.92)

3. (a) 7

   (b) 21

4. (a) $\frac{1}{2}$

   (b) $\frac{1}{6}$

5. $\frac{17}{40}$

6. $5\frac{9}{10}$

7. (a) $\frac{1}{2}$

   (b) $4\frac{5}{12}$

8. 40

1F - Decimals and percentages

1. 0.07, 0.072, 0.7, 0.702, 0.72

2. £238

3. £30

4. (a) £320

   (b) £5.60

5. 60%

6. (a) 85%

   (b) 10%

   (c) 0.6

7. (a) $\frac{2}{5}$

   (b) 0.98

8. $\frac{63}{100}$

9. $\frac{1}{5}$, 22%, $\frac{2}{7}$, 0.3,

10. £14 734.50

11. £78

12. £58.75

1G - Money

1. £12.60

2. £5.63

3. (a) £5.55

   (b) 21

4. 60

5. £72.96

6. (a) 197

   (b) £41.37

7. 64p

8. £2.38

9. (a) £2.40

   (b) £2.00

   (c) 1.25 €

10. (a) USA

    (b) £1.99 or $ 2.83

1H - Ratio and proportion

1. (a) $2 : 3$

    (b) 20

2. 62.5cm

3. (i) $\dfrac{2}{7}$

    (ii) 12

4. 18
    2100ml
    195g

1J - Estimations

1. 3000 - 3750

2. $20 - 21\dfrac{1}{3}$

2A - Introducing algebra

1. (a) $3x$

    (b) $x - 9$

2. (a) $7p + 3q$

    (b) $\dfrac{180 - x}{2}$

3. $8x + 8(x + 2)$

4. 6

5. (a) $6m$

    (b) $4p + 5q$

    (c) $3t - 12$

    (d) $12\,ab$

6. (a) $2p + 4q$

    (b) $2y^2$

    (c) $3c + 4d$

    (d) $8pq$

7. (i) $9c$ (ii) $3fg$ (iii) $3x^2$

8. $2f + 7g$

9. (a) (i) $a + 3b$

    (ii) $2x^2 + x$

    (b) (i) $8x - 12$

    (ii) $pq - p^3$

    (c) $5p + 16$

10. (a) $5a + 2b$

    (b) $2x^3$

11. (a) $3t - 12$

    (b) $6\,(q + 2)$

12. $y^4 + 2y^2$

13. $9y + 5$

14. p^6

2B - Expanding and factorising quadratics

1. $y^2 + 7y + 12$

2. $x^2 + 3x - 28$

3. $x(x - 3)$

4. (a) $p(p + 6)$

   (b) $3x(2x - 3y)$

5. $2p(p - 2q)$

2C - Formulae

1. (a) £51
   (b) £60

2. (a) £70
   (b) 9

3. 219°C

4. 22

5. (a) £57.60
   (b) £291.60

6. $12x - 3$

7. -25

8. (a) -5
   (b) 73

9. 40

10. $C = 10(n + 3)$

11. (a) $T = 4n + 3c$
    (b) 5

12. $P = 5x + 2y$

13. $\dfrac{ac + d}{a}$ or $\dfrac{d}{a} + c$

14. $t = \dfrac{v - u}{5}$

15. $h = \dfrac{f - g}{3}$

2D - Equations and inequalities

1. (a) $4x + 8$

   (b) 15.5 cm

2. 14

3. (a) $x + 10 + 2x + x + 90 + x + 20 = 360$

   $5x + 120 = 360$

   (b) $58°$

4. $(3x - 2)(2x + 5) + 2(3x - 2) = 25$

5. 5

6. (a) 2

   (b) 0.5

7. -1.5

8. $-\dfrac{7}{3}$ or $-2\dfrac{1}{3}$

9. (a) $\dfrac{3}{5}$

   (b) $-3, -2, -1, 0, 1, 2$

10. $y = -1.5$

11. (a) -3
    (b) 1.4
    (c) -3
    (d) 3.5

12. 1.7

13. 3.3

14. $x = -2, y = -2$
    $x = 0, y = 3$

15. (i) $x > \dfrac{20}{7}$ (ii) 3

3A - Sequences

1. (a)

   (b) 21

2. (a) (i) 5

      (ii) 23

   (b) $\times 2 - 1$

   (c) •••••
       ••••••
       ••••

   (d) 14, 17

3. (a) ⌂⌂⌂⌂

   (b) 60

   (c) $m = 6n$

4. (a) 106 102

   (b) take away 4 each time

   (c) 46

5. (a) 40

   (b) it is not an even number

6. (a) 27

   (b) 9 13 17

7. (a) $4n + 3$

   (b) $453 - 3$ is not a multiple of 4

3B - Real life graphs

1. (a) $(-2, 2)$

    (b)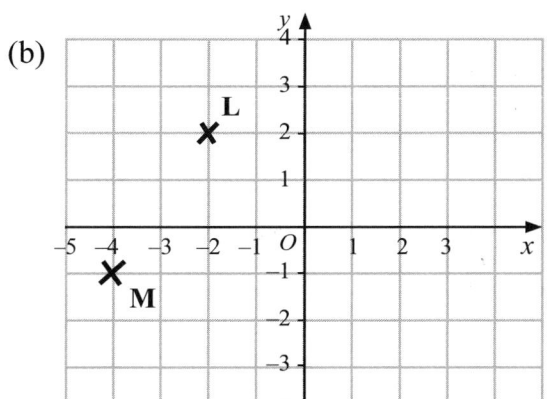

2. (i) £80

    (ii) 125 miles

3. (a) 15 minutes

    (b) 15 km

    (c)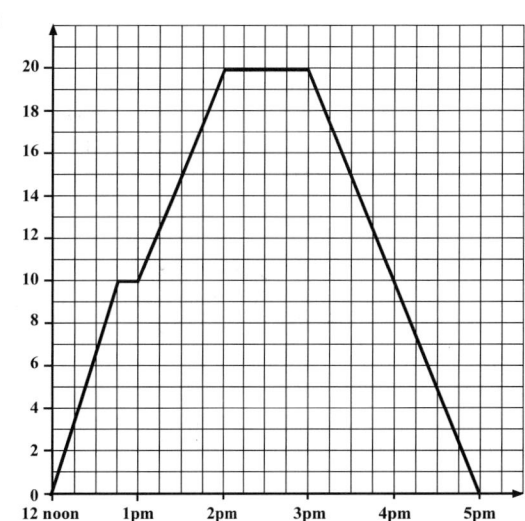

4. (a) 10 kg

    (b) 5.5 pounds

    (c) 50 kg

3C - Graphs on coordinate axes

1. (a) $-2, 4, 7, 10$

    (b)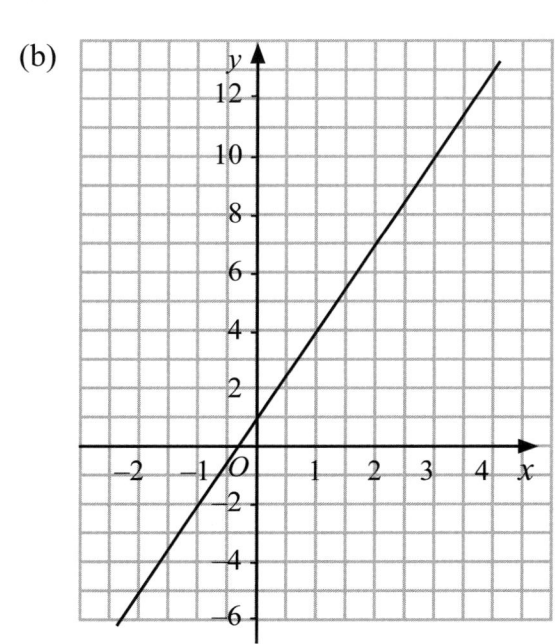

    (c) (i) -1.4

         (ii) 2.4

2. (a)

x	-2	-1	0	1	2
y	**-4**	-1	**2**	5	**8**

    (b)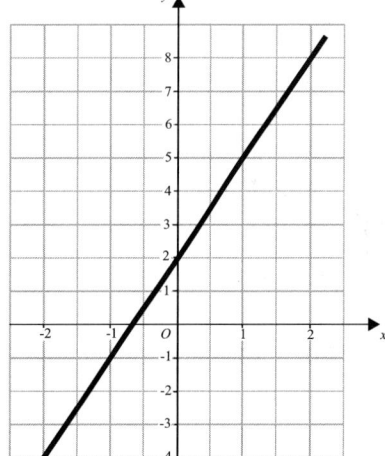

3. (a) 6 2 **0** **0** **2** 6 **12**

    (b)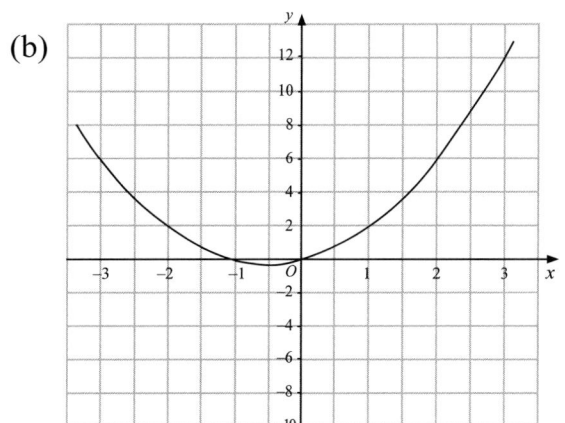

4. $-0.6, 2.6$

A03

Shape, space and measures

Contents

4 Geometrical reasoning

4A - Angles
LIN 3, 11 MOD 7, 14

- ❏ Name and recognise angles
- ❏ Distinguish between alternate and corresponding angles on parallel lines
- ❏ Know the angle properties of a parallelogram and a rhombus
- ❏ Find a missing angle using angles on a straight line = 180° and angles at a point = 360°

- ❏ Find angles in a triangle knowing that their sum is 180°, exterior angles = sum of interior opposites and base angles of an isosceles triangle are equal
- ❏ Find the exterior angle using the sum of the exterior angles of a polygon is 360°
- ❏ Find the size of interior and exterior angles of a regular polygon

1. This triangle is accurately drawn.

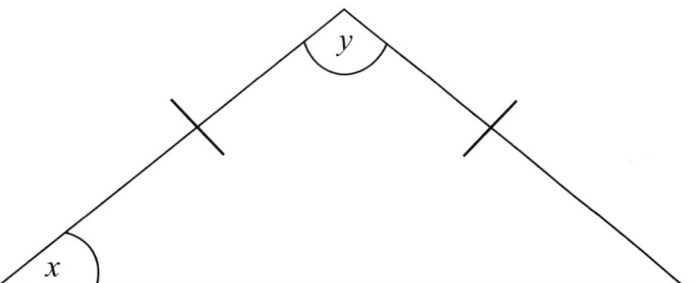

(a) Write down the special name for this type of triangle.

....

(1)

(b) What type of angle is angle x?

....

(1)

(c) What type of angle is angle y?

....

(1)

2.

Diagram **NOT** accurately drawn

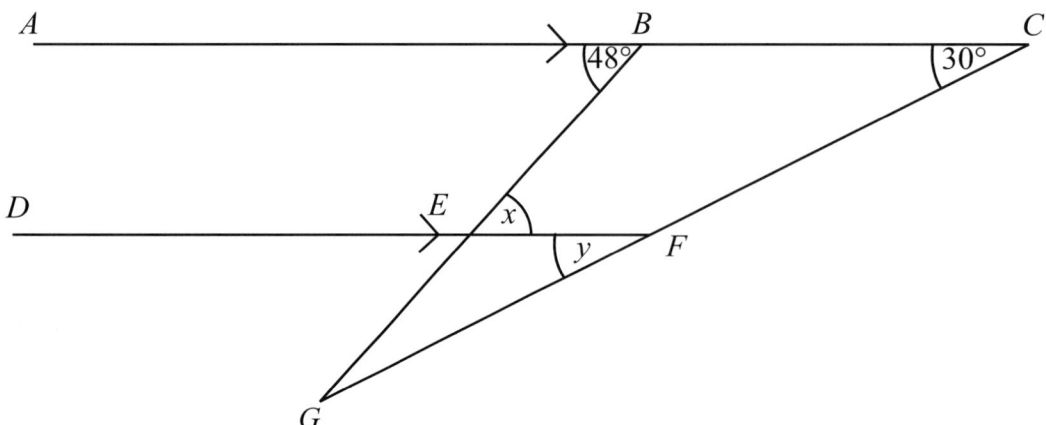

BEG and *CFG* are straight lines.
ABC is parallel to *DEF*.
Angle *ABE* = 48°.
Angle *BCF* = 30°.

(a) (i) Write down the size of the angle marked *x*.

x = ...°

    (ii) Give a reason for your answer.

....

(2)

(b) (i) Write down the size of the angle marked *y*.

y = ...°

    (ii) Give a reason for your answer.

....

(2)

3.

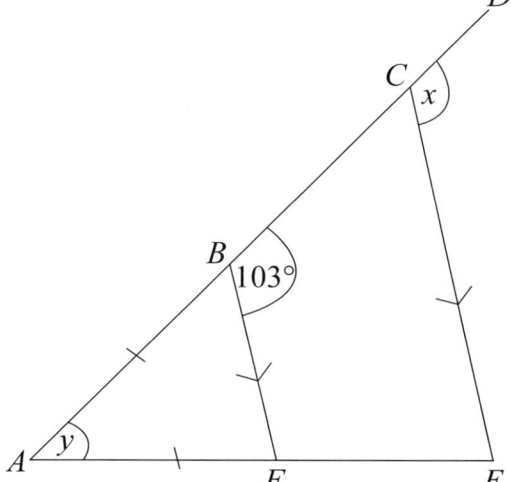

Diagram **NOT** accurately drawn

ABCD and *AFE* are straight lines.
BF is parallel to *CE*.
Angle *CBF* = 103°.
AB = *AF*.

(a) (i)   Find the size of angle *x*.

.....................................°

(ii) Give a reason for your answer.

....

....

(2)

(b) Find the size of angle *y*.

.....................................°

(2)

A03 - Shape, space and measures

4.

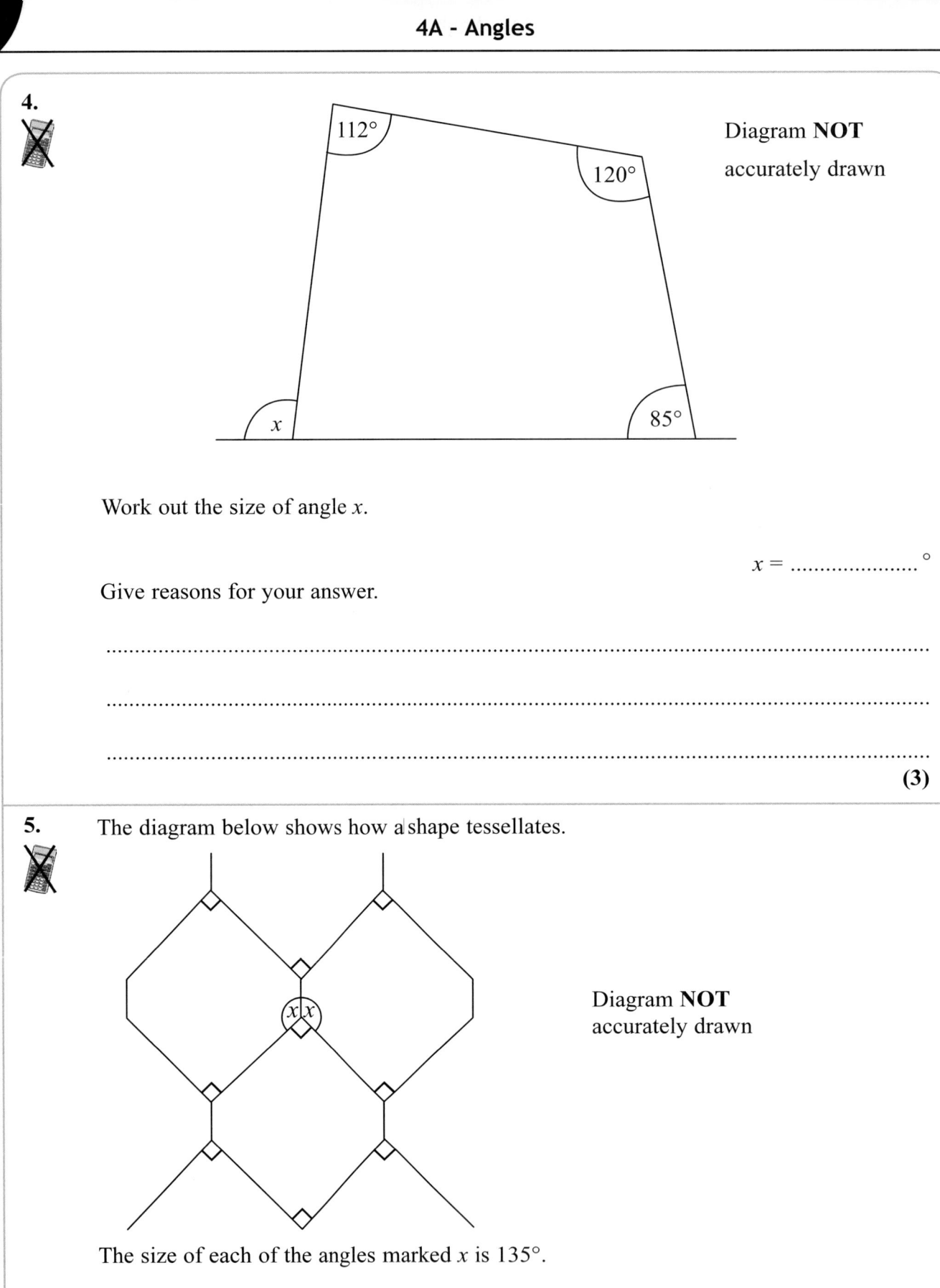

Diagram **NOT** accurately drawn

Work out the size of angle *x*.

$x =$ .... °

Give reasons for your answer.

....

....

....

(3)

5. The diagram below shows how a shape tessellates.

Diagram **NOT** accurately drawn

The size of each of the angles marked *x* is 135°.

Give reasons why.

....

....

....

(2)

6.

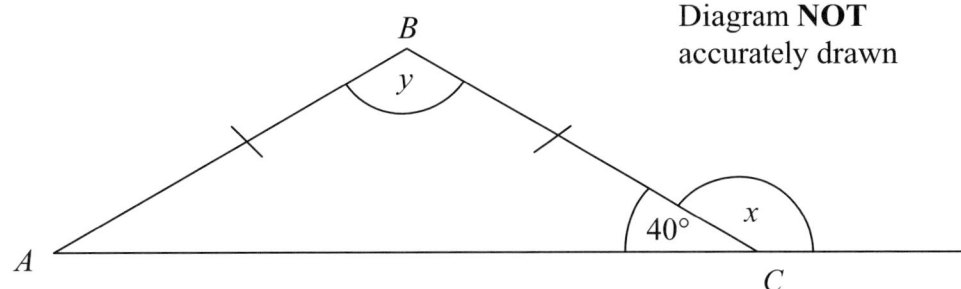

Diagram **NOT** accurately drawn

In triangle ABC,
$AB = BC$,
Angle $ACB = 40°$

(i)   Work out the size of angle x.

.... °

(ii) Give a reason for your answer.

....

....

(2)

A03 - Shape, space and measures

7.

The diagram shows a 5-sided shape.
All the sides of the shape are equal in length.

Diagram **NOT**
accurately drawn

(a) (i) Find the value of *x*.

$x = $ ....

(ii) Give a reason for your answer.

....

(2)

(b) Work out the value of *y*.

$y = $ ....

(2)

8.

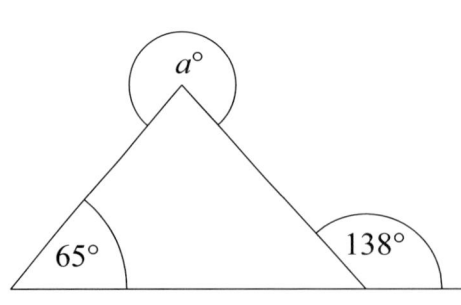

Diagram **NOT**
accurately drawn

Work out the value of *a*.

$a = $ ....

(3)

9.

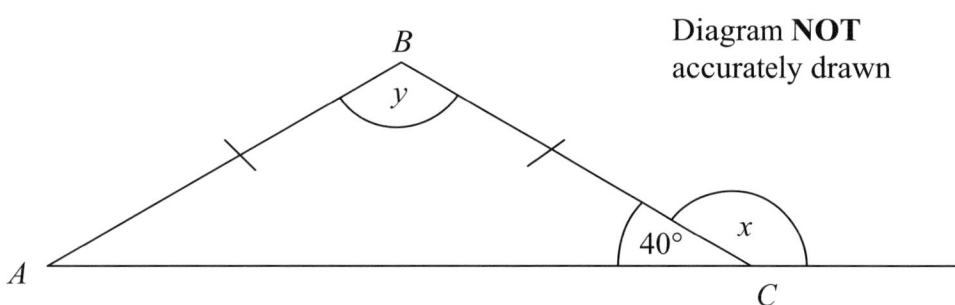

Diagram **NOT** accurately drawn

In triangle *ABC*,
AB = *BC*,
Angle *ACB* = 40°

(i) Work out the size of angle *y*.

.... °

(ii) Give a reason for your answer.

....

....

(3)

10.

The diagram shows the exterior angles of a quadrilateral.

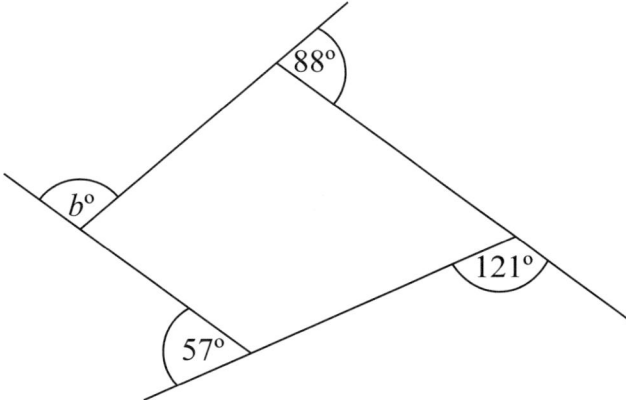

Diagram **NOT** accurately drawn

Work out the value of *b*.

b = ...

(2)

11.

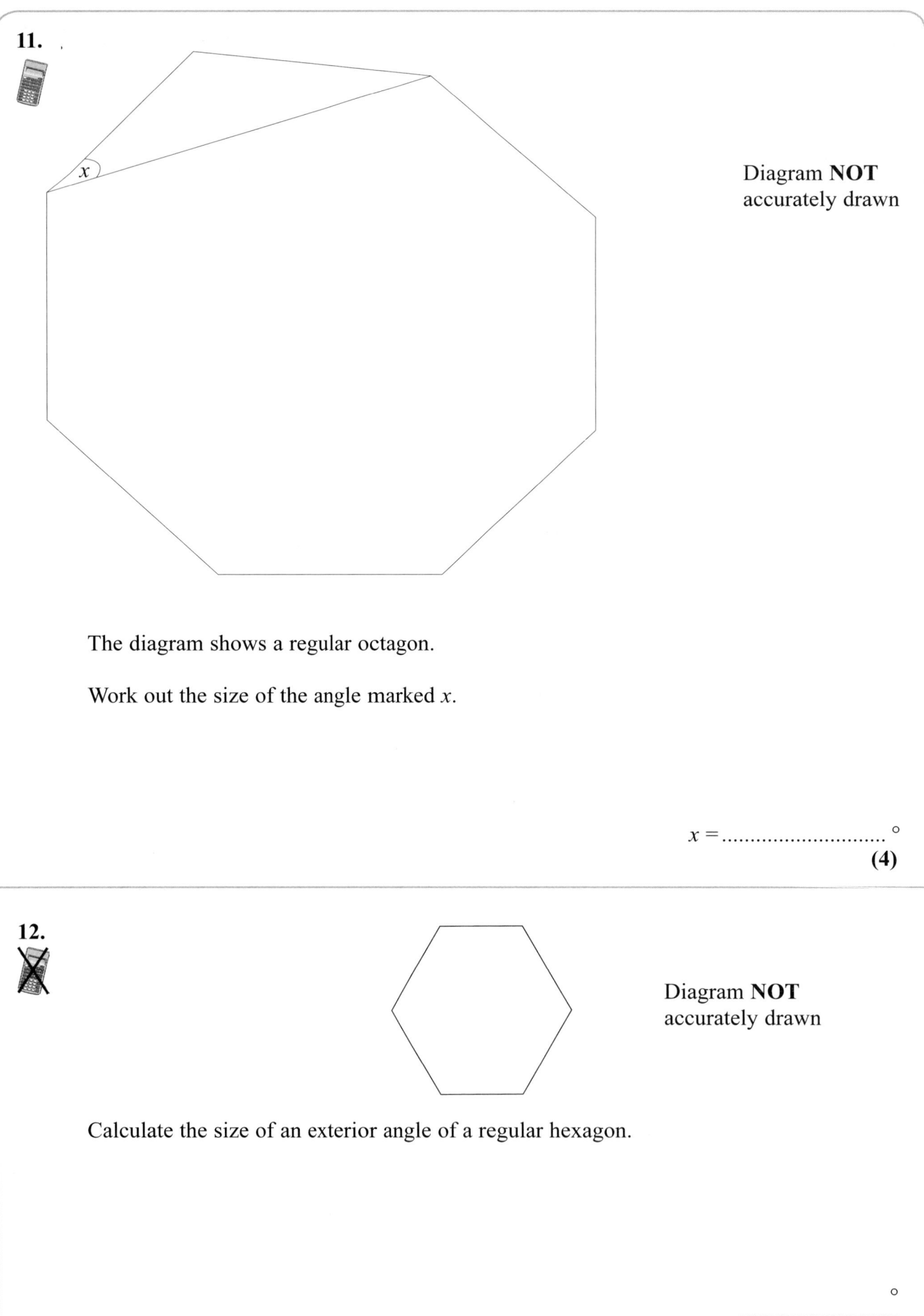

Diagram **NOT** accurately drawn

The diagram shows a regular octagon.

Work out the size of the angle marked x.

$x =$ .... $^\circ$

(4)

12.

Diagram **NOT** accurately drawn

Calculate the size of an exterior angle of a regular hexagon.

.... $^\circ$

(2)

4B - Properties of Shapes *LIN 5 MOD 23*

❏ Name and recognise all types of triangle ❏ Name, sketch and draw parts of a circle

❏ Know that congruent shapes are identical ❏ Show how shapes tessellate by drawing

❏ Name, recognise, sketch and draw quadrilaterals

1.

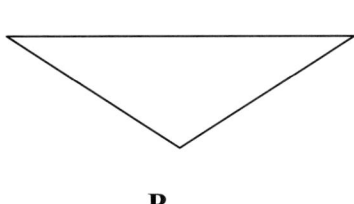

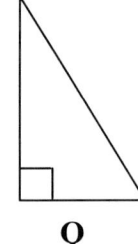

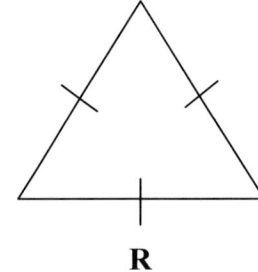

P   **Q**   **R**

(i)   Draw a line of symmetry on triangle **P**.

(ii) Write down the mathematical name for triangle **Q**.

.... triangle

(iii) Write down the mathematical name for triangle **R**.

.... triangle

(3)

2.   This triangle is accurately drawn.

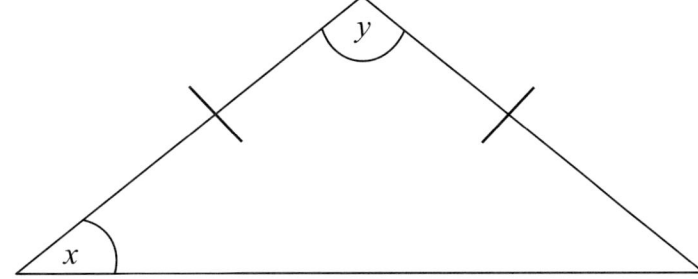

(a) Write down the special name for this type of triangle.

....

(1)

(b) What type of angle is angle *x*?

....

(1)

(c) What type of angle is angle *y*?

....

(1)

A03 - Shape, space and measures

3. Here are 8 shapes on a grid of centimetre squares.

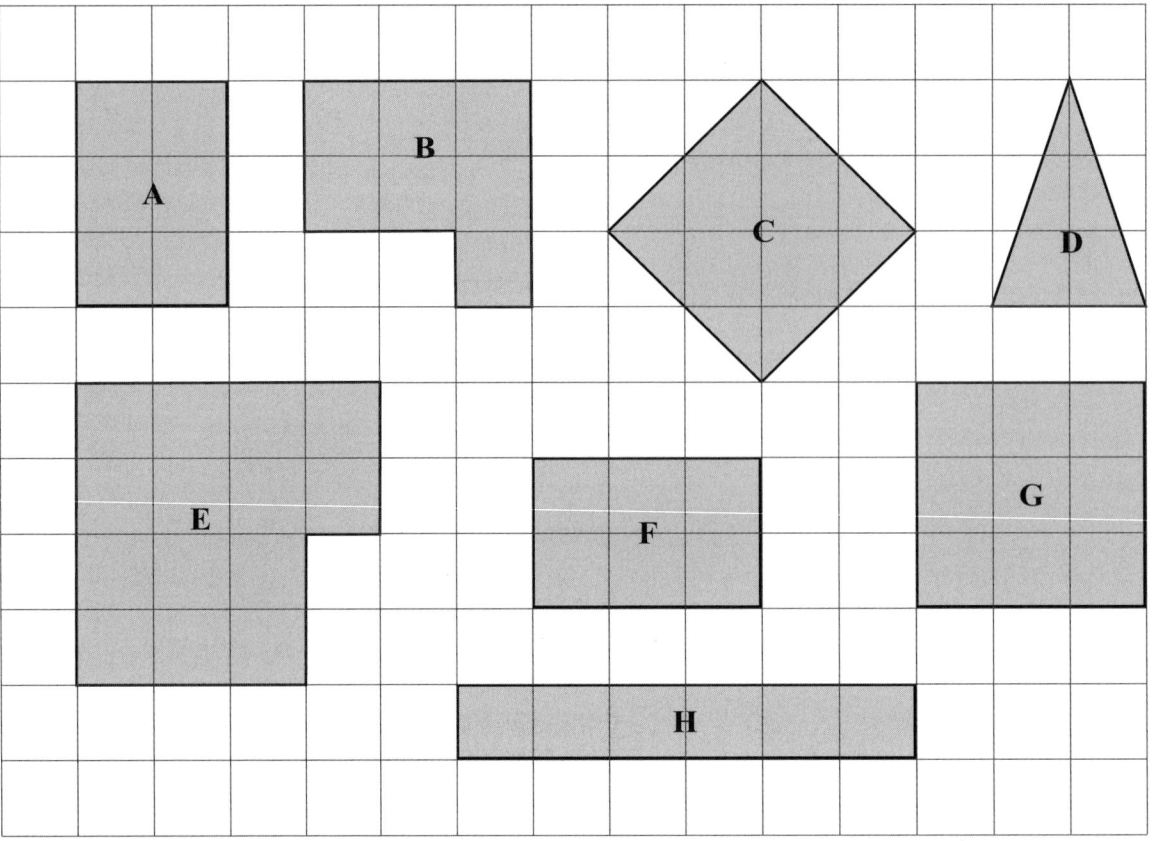

(a) Write down the special name for triangle **D**.

....

(1)

Two of the shapes are congruent.

(b) Write down the letters of these two shapes.

.... and ....

(1)

4. **Two** of these triangles are congruent.

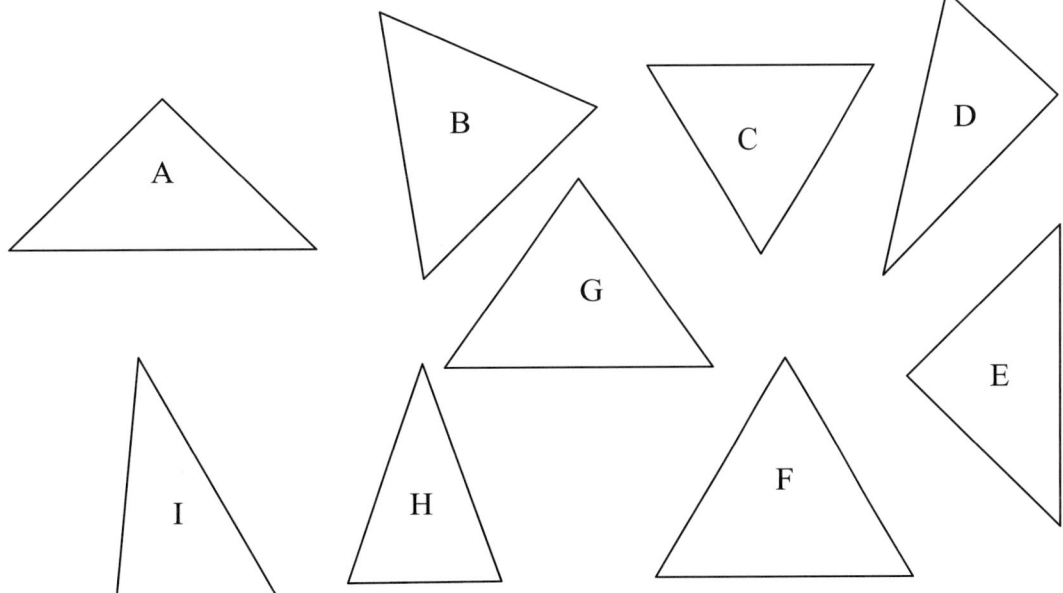

(a) Write down the letters of the two triangles that are congruent.

........................., ...

(1)

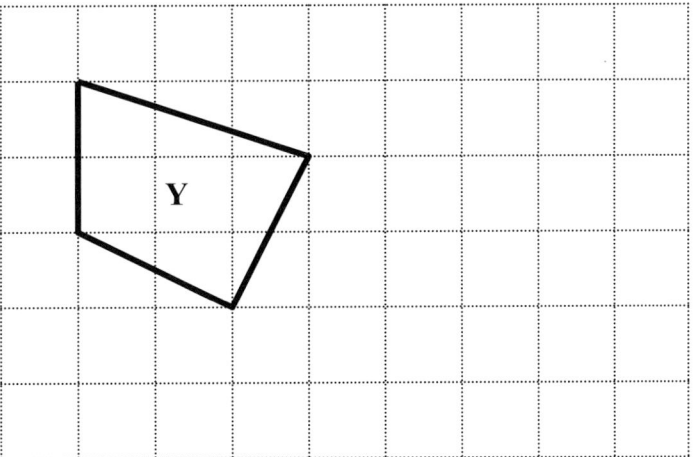

(b) On the grid draw a shape that is congruent to shape **Y**.

(1)

A03 - Shape, space and measures

5.

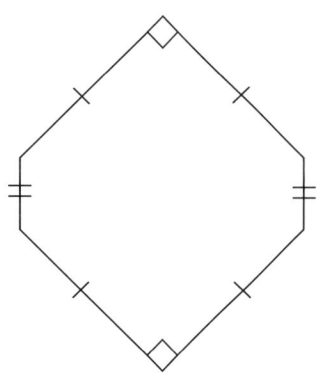

Diagram **NOT**
accurately drawn

The diagram shows a shape.
The shape is a 6-sided polygon.

Write down the mathematical name for a 6-sided polygon.

....

(1)

6. Here are some diagrams relating to a circle. (3)

Draw an arrow from each of the diagrams to its mathematical name.
The arrow showing an arc is drawn for you.

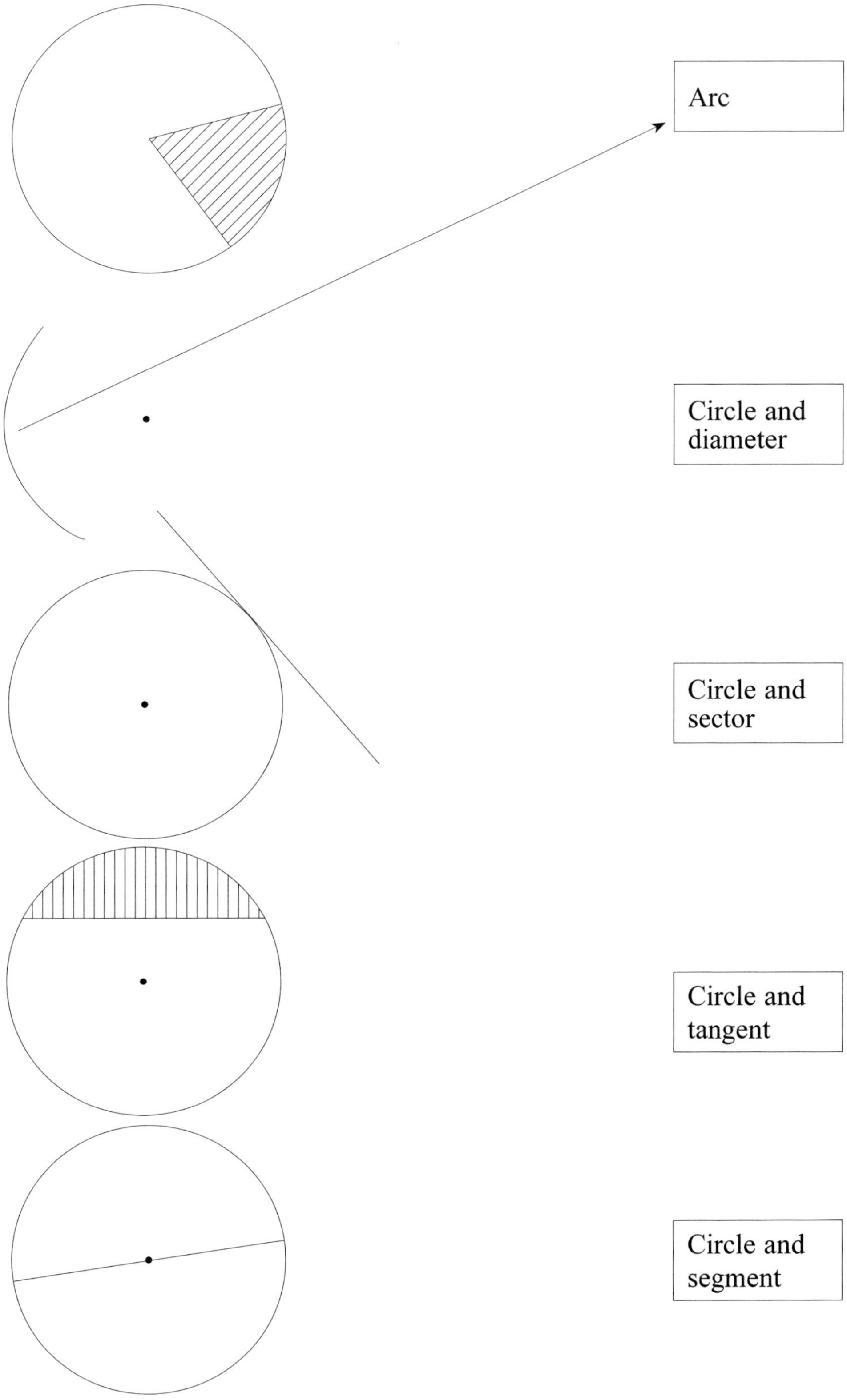

| Arc |

| Circle and diameter |

| Circle and sector |

| Circle and tangent |

| Circle and segment |

(3)

A03 - Shape, space and measures

7. On the grid below, show how the shaded shape will tessellate.

You should draw at least six shapes.

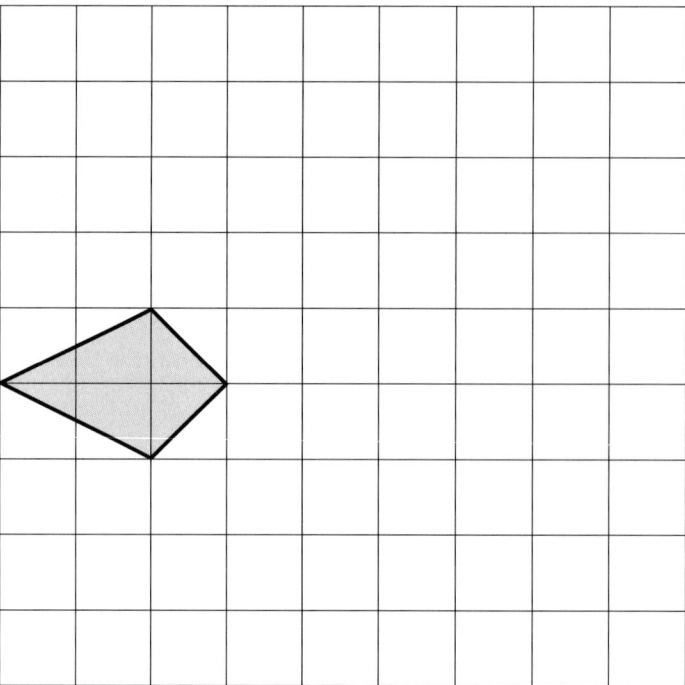

(2)

8. The diagram shows a trapezium on a grid.

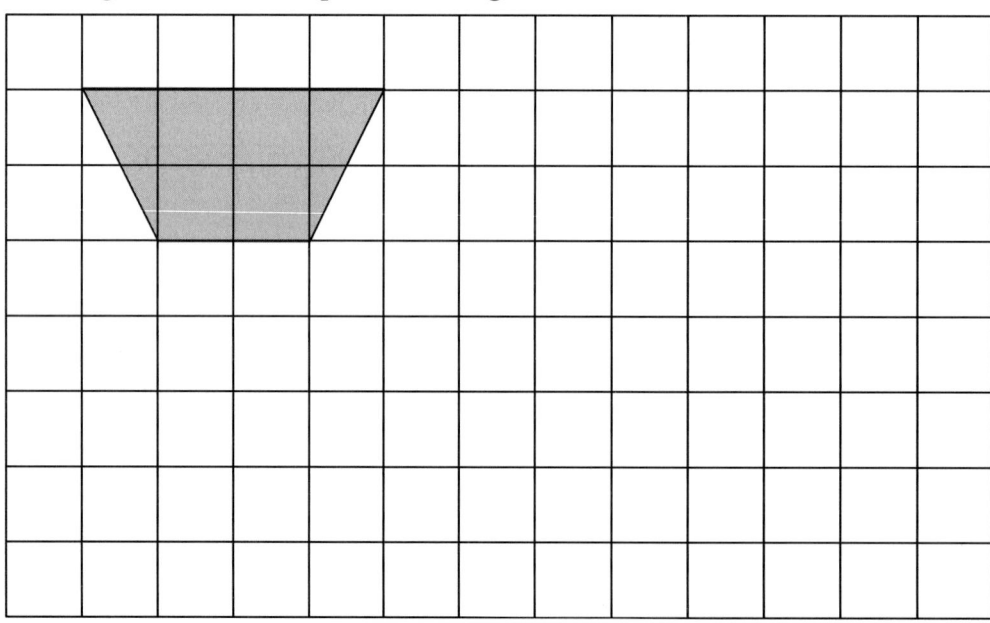

Show how the trapezium tessellates.

You should draw at least 6 shapes on the grid.

(2)

4C - 3-D shapes and volumes

LIN 25 MOD 20

❏ Identify faces, edges and vertices of 3-D shapes

❏ Name and sketch solids

❏ Match nets to solids

❏ Draw nets of shapes and use them

❏ Draw and use plan and elevation of 3-D shapes

❏ Find the surface area of shapes by using area of known 2-D shapes

❏ Find the volume of simple shapes by counting cubes and part cubes

❏ Calculations volumes of cuboids and use inverse calculations

❏ Calculate volumes of prisms and cylinders

1. Write down the mathematical name of each of these two 3-D shapes.

(i)

(ii)

(i).......................................

(ii).......................................

(2)

2.   The diagram shows some solid shapes and their nets.
An arrow has been drawn from one solid shape to its net.

Draw an arrow from each of the other solid shapes to its net.

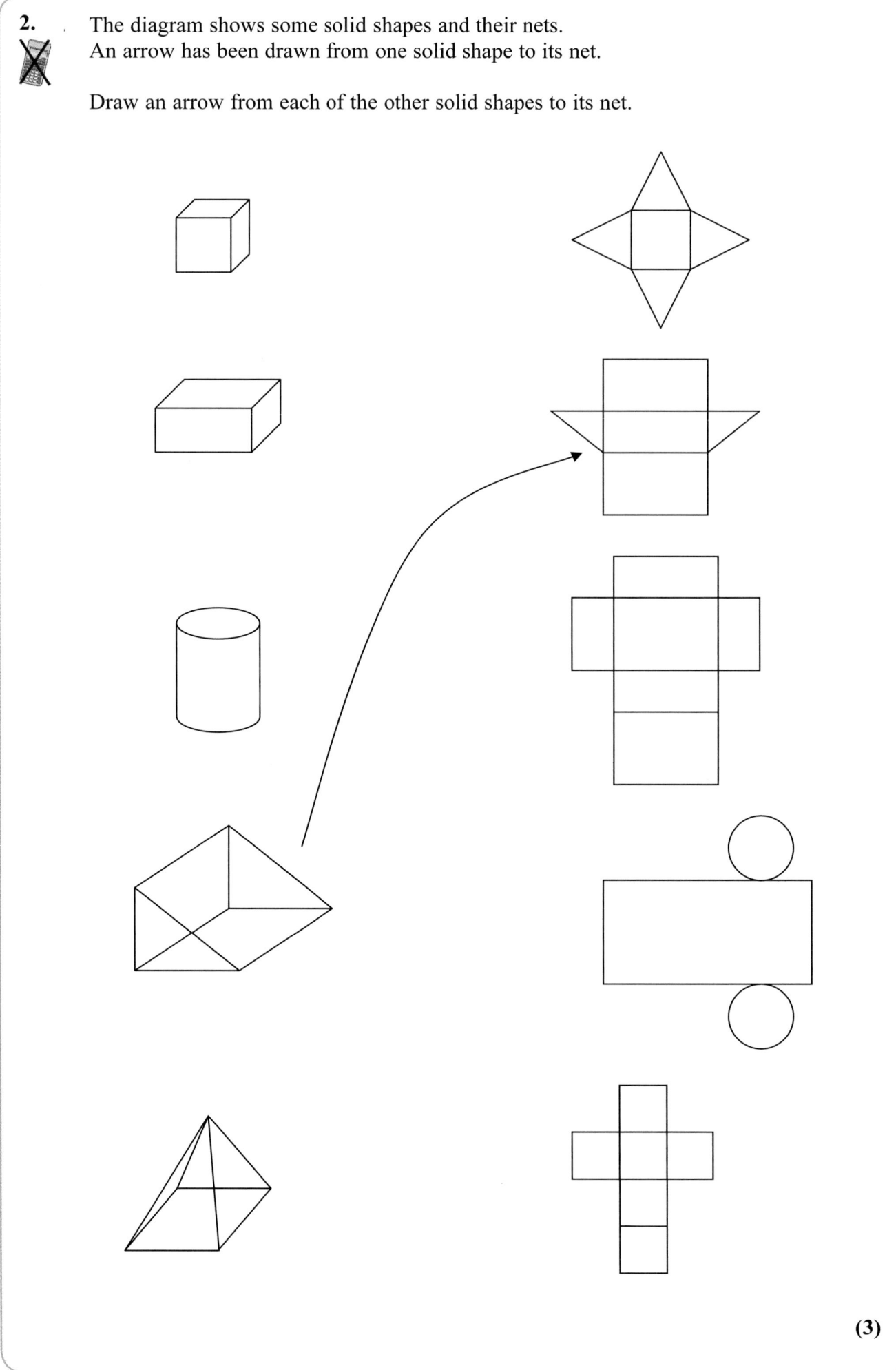

(3)

3.

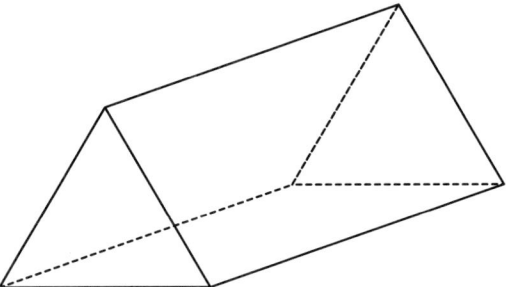

The diagram shows a triangular prism.
The cross-section of the prism is an equilateral triangle.

In the space below, draw a sketch of a net for the triangular prism.

(2)

4. Here are the plan, front elevation and side elevation of a 3-D shape.

plan

**front
elevation**

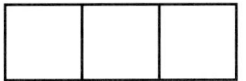

**side
elevation**

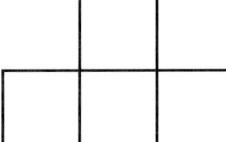

In the space below, draw a sketch of the 3-D shape.

(2)

A03 - Shape, space and measures

5.

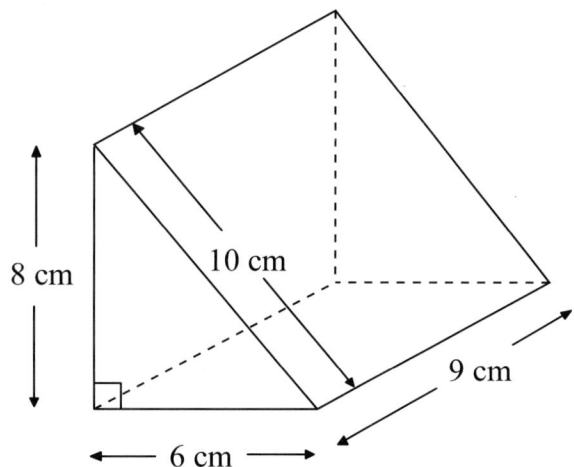

8 cm

10 cm

9 cm

6 cm

Diagram **NOT**
accurately drawn

Work out the surface area of the triangular prism.
State the units with your answer.

(4)

6. Find the volume of this prism.

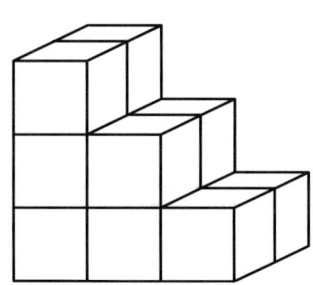

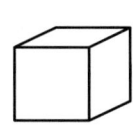

Diagram **NOT**
accurately drawn

represents 1 cm³

....................cm³

(2)

7. Here is a solid shape.

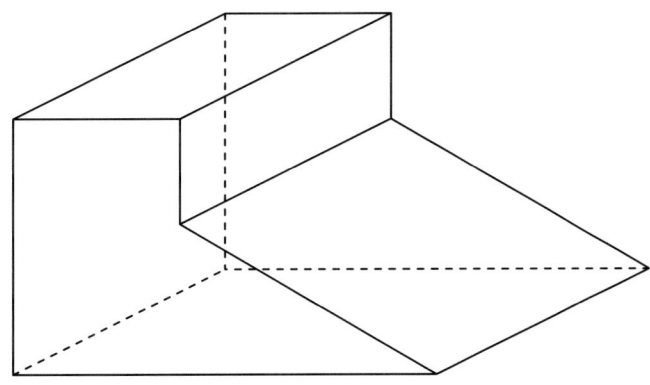

Diagram **NOT**
accurately drawn

Write down the number of;

(i) faces,

.... faces

(ii) edges,

.... edges

(iii) vertices.

.... vertices

(3)

8.

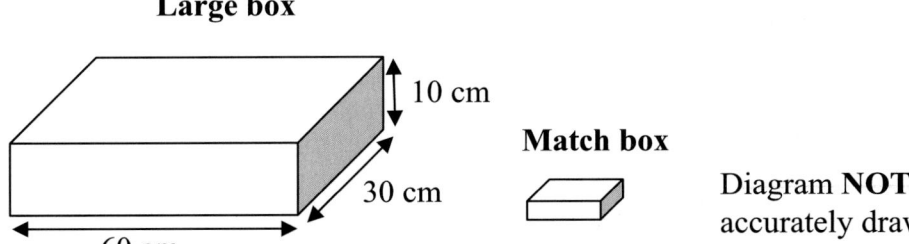

Large box

10 cm

Match box

Diagram **NOT**
accurately drawn

30 cm

60 cm

The diagram shows a large box in the shape of a cuboid and a matchbox.

The large box is full of match boxes.
Each match box is in the shape of a cuboid.
Each match box is 6 cm by 3 cm by 1 cm.

Work out the number of match boxes in the large box.

....

(3)

A03 - Shape, space and measures

9.   A cuboid has   **(3)**

a volume of 40 cm^3
a length of 5 cm
a width of 2 cm

(a) Work out the height of the cuboid.

.......................cm
(2)

(b)

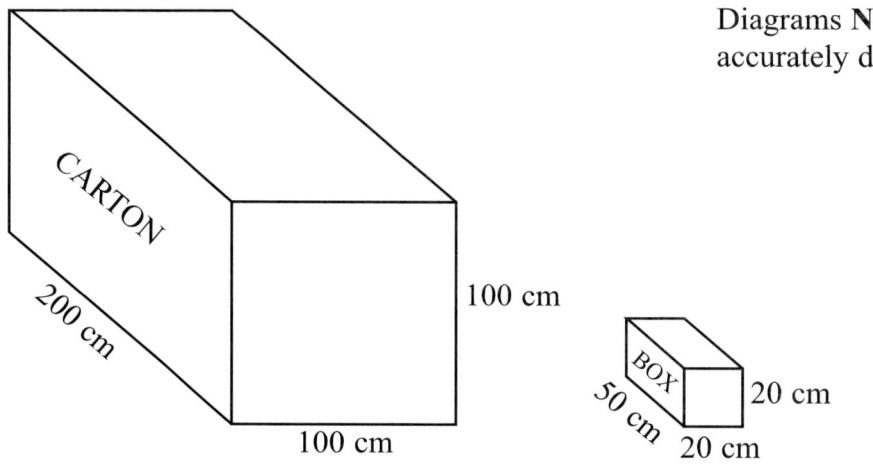

Diagrams **NOT**
accurately drawn

A carton measures 200cm by 100cm by 100cm.

The carton is to be completely filled with boxes.

Each box measures 50cm by 20cm by 20cm.

Work out the number of boxes which can completely fill the carton.

....
(3)

4D - Pythagoras   *LIN 31 MOD 38*

❏ Use Pythagoras to find one side of a right angle triangle given the other two sides

1.

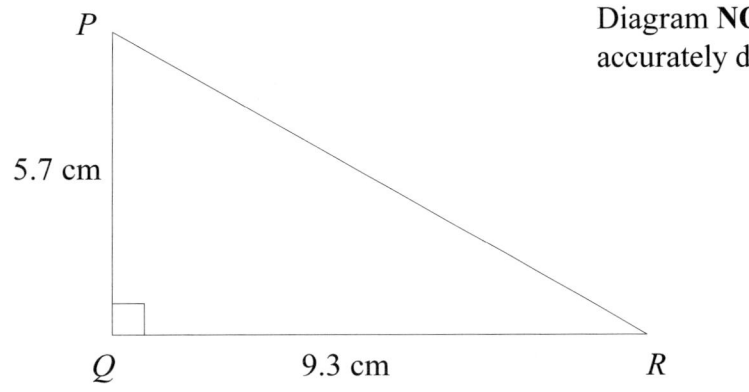

Diagram **NOT** accurately drawn

In triangle *PQR*, *QR* = 9.3 cm.   *PQ* = 5.7 cm.   Angle *PQR* = 90°.

Calculate the length of *PR*.
Give your answer correct to 3 significant figures.

.... cm

(3)

2.

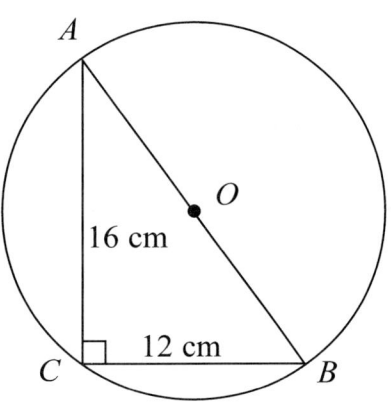

Diagram **NOT** accurately drawn

The diagram shows triangle *ABC* and a circle, centre *O*.
A, *B* and *C* are points on the circumference of the circle.
AB is a diameter of the circle.

AC = 16 cm and *BC* = 12 cm.

Work out the diameter *AB* of the circle.

.... cm

(3)

3.

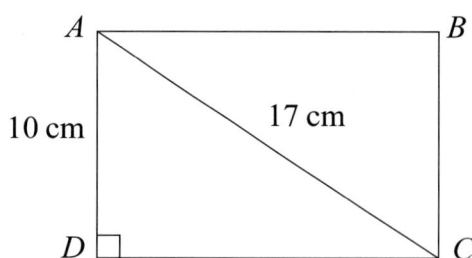

Diagram **NOT**
accurately drawn

ABCD is a rectangle.
AC = 17 cm.
AD = 10 cm.

Calculate the length of the side *CD*.
Give your answer correct to one decimal place.

......................cm

(3)

4.

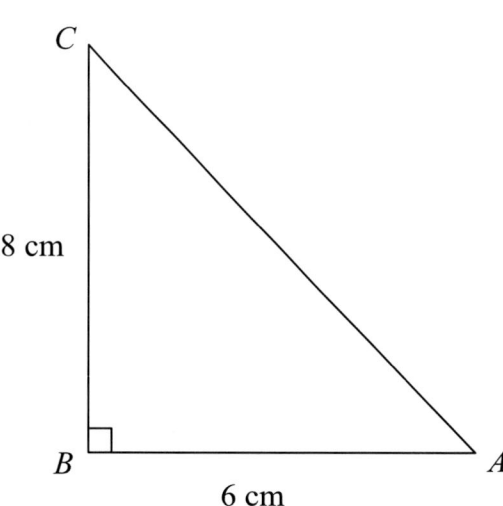

Diagram **NOT**
accurately drawn

Calculate the length of *AC*.

.... cm

(3)

5 Transformation and coordinates

- ❏ Draw a line of symmetry
- ❏ Recognise reflective symmetry
- ❏ Identify all lines of symmetry in a shape
- ❏ Add one square to complete a shape with 1 line of symmetry

- ❏ State order of symmetry and pick out shapes with a stated order
- ❏ Add 1 square to complete a shape with order 2
- ❏ Identify rotational symmetry

1.

mirror line

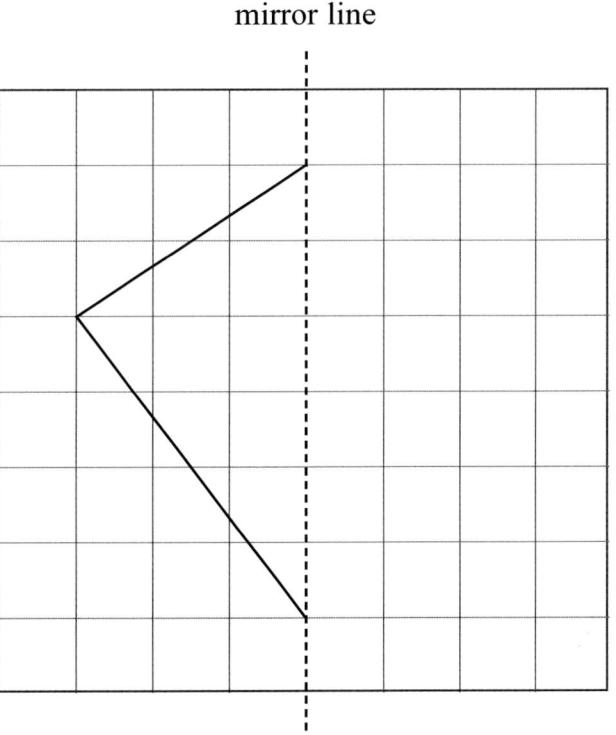

(i) Draw the reflection of the shape above in the mirror line.

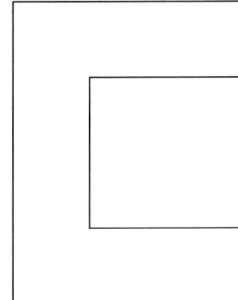

(ii) Draw in the line of symmetry of this shape.

(2)

A03 - Shape, space and measures

2. A shaded shape has been drawn on the centimetre grid.

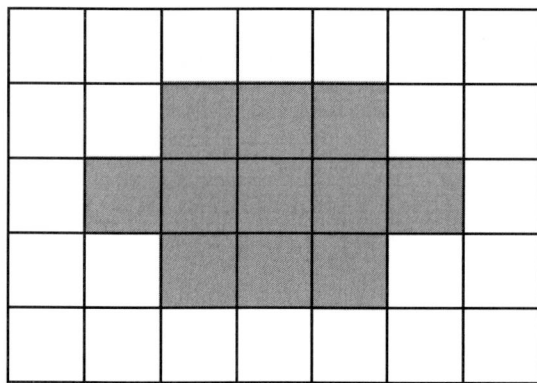

The shaded shape has **two** lines of symmetry.

Draw the **two** lines of symmetry on the shaded shape.

(2)

3.

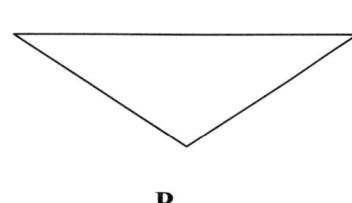   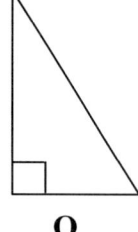   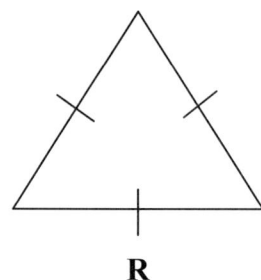

     P   **Q**   **R**

(i)   Draw a line of symmetry on triangle **P**.

(ii) Write down the mathematical name for triangle **Q**.

.... triangle

(iii) Write down the mathematical name for triangle **R**.

.... triangle

(3)

4.

On the star draw in all the lines of symmetry.

(1)

5. . (a) On the grid below, 6 squares are shaded.

Shade one more square so that the shaded shape has one line of symmetry.

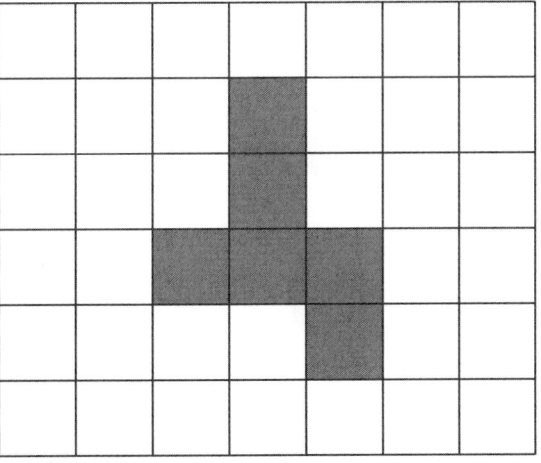

(1)

(b) On the grid below, 4 squares are shaded.

Shade one more square so that the shaded shape has rotational symmetry of order 2.

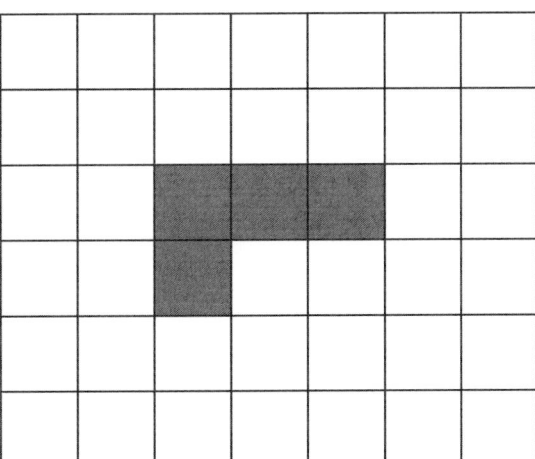

(1)

A03 - Shape, space and measures

6. Here are three shapes.

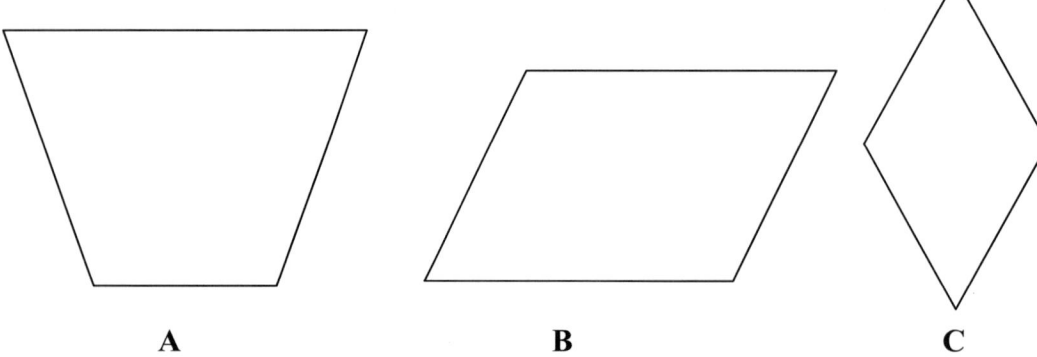

A   B   C

One of these shapes has **no** lines of symmetry.

Write down the letter of this shape.

....

(1)

7.

11   16   18   36

68   69   82   88

From these numbers, write down a number which has

(i)   exactly **one** line of symmetry,

....

(ii)   2 lines of symmetry **and** rotational symmetry of order 2,

....

(iii) rotational symmetry of order 2 but **no** lines of symmetry.

....

(3)

8. The diagram shows part of a shape.

The shape has rotational symmetry of order 4 about the point *P*.

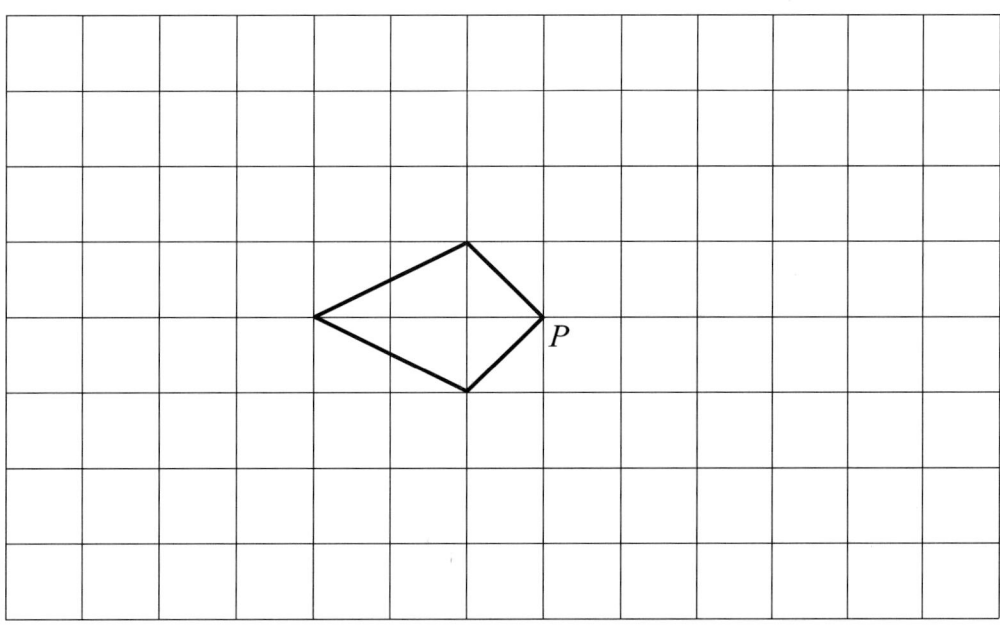

On the grid above, complete the shape.

(3)

9. Write down the order of rotational symmetry of this star.

order ....

(1)

A03 - Shape, space and measures

5B - Coordinates

❏ Identify and plot coordinates in all 4 quadrants

❏ Find the coordinates of midpoints of lines
   (diagram given)

❏ Find the coordinates of midpoints between
   given points (no diagram given)

1.

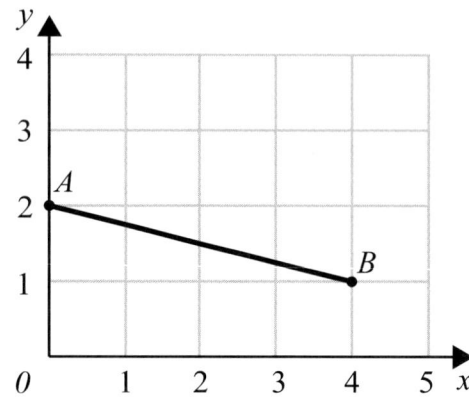

(a) Write down the coordinates of the point

   (i) A,

                                                   (……… , ……..)

   (ii) B.

                                                   (……… , ……..)

                                                        (2)

(b) On the grid, mark with a cross (×) the midpoint of the line AB.

                                                                        (1)

2.

A is the point $(4, 3)$

B is the point $(-2, 1)$

Find the coordinates of the midpoint of the line AB.

                                                  (……….. , ………..)

                                                                  (2)

5C - Transformations

LIN 26 MOD 31

❏ Reflect shapes on coordinate axes

❏ Reflect shapes in 2 dimensions

❏ Rotate a shape on a grid

❏ Translate of a shape using vectors

❏ Enlarge a shape by a given scale factor

❏ Describe, in words, a transformation or a combination of transformations

1.

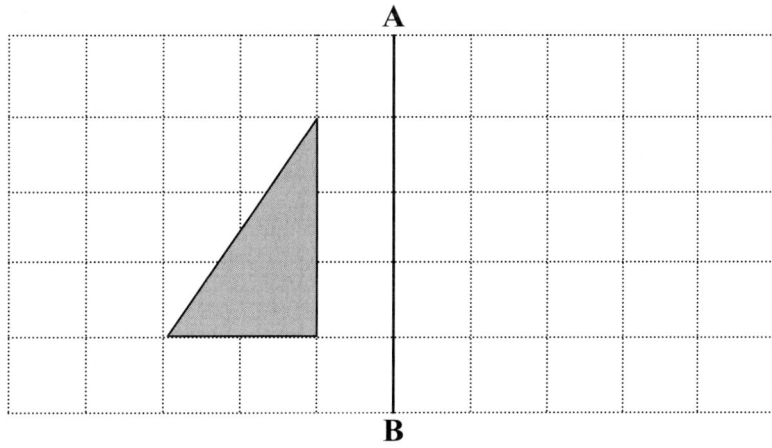

Reflect the shaded triangle in the line **AB**.

(1)

2.

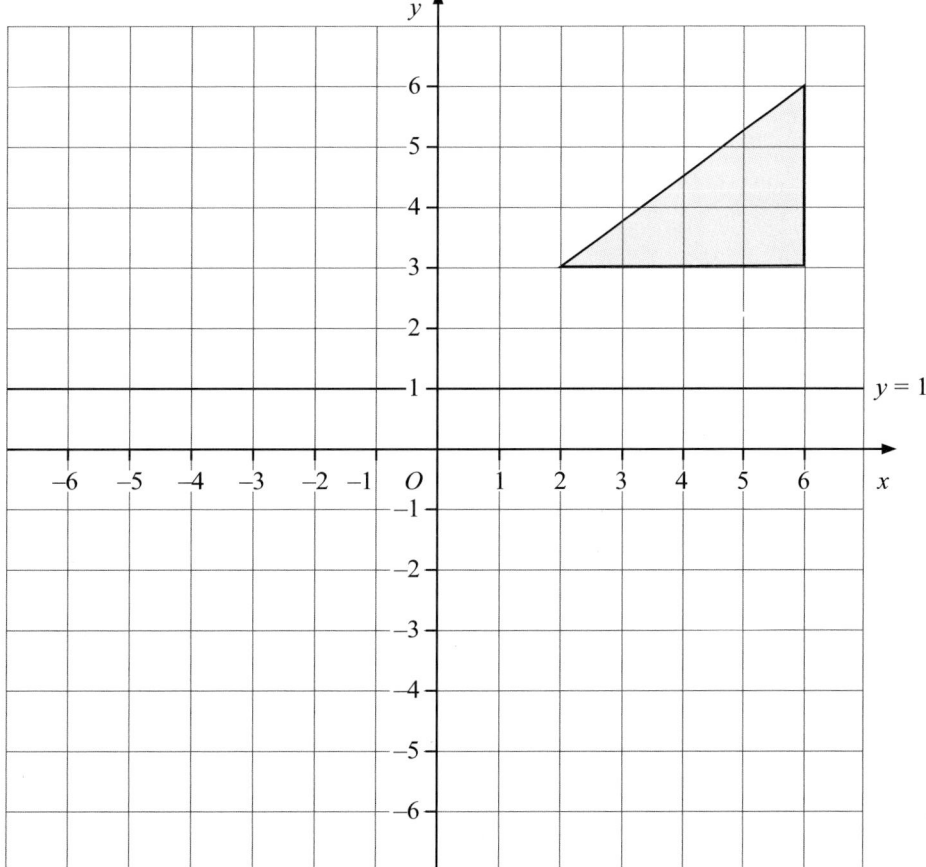

Reflect the triangle in the line $y = 1$

(2)

A03 - Shape, space and measures

3.

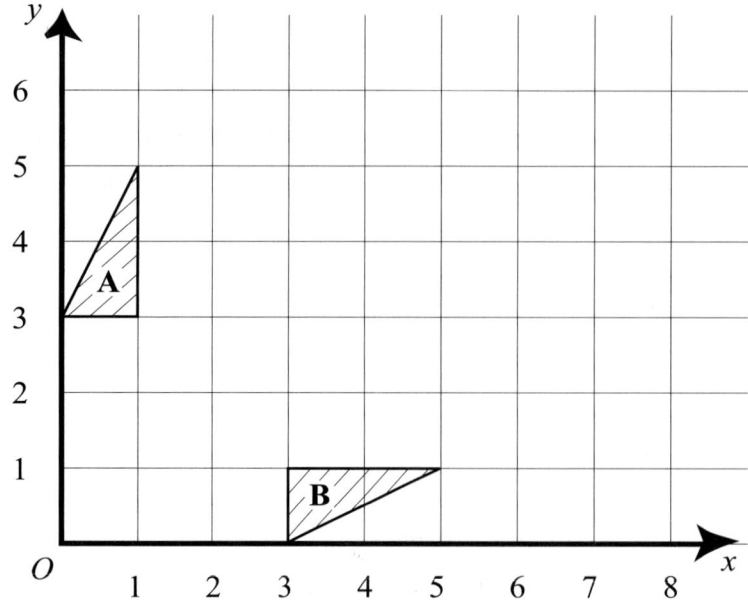

Triangle **A** and triangle **B** have been drawn on the grid.

(a) Reflect triangle **A** in the line $x = 3$.
Label this image **C**.

(2)

(b) Describe fully the single transformation which will map triangle **A** onto triangle **B**.

....

(2)

4.

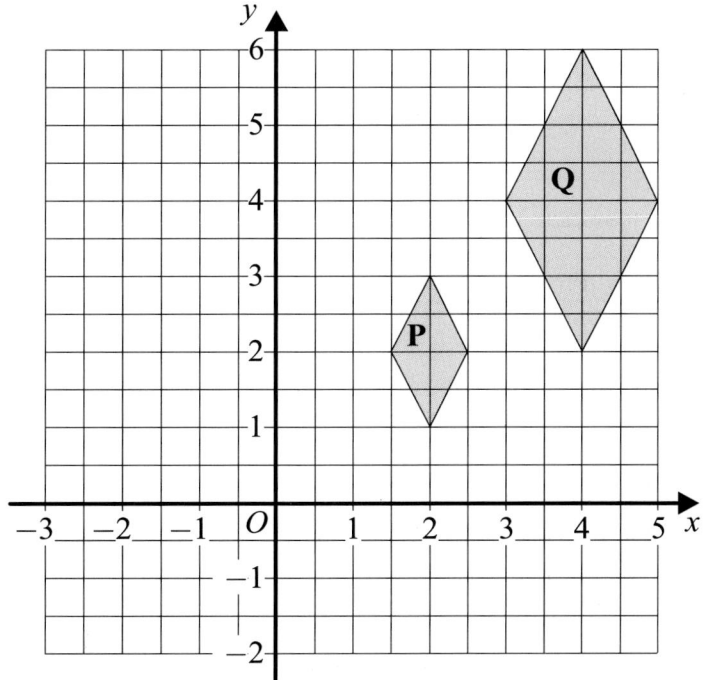

(i) Describe fully the single transformation that maps shape **P** onto shape **Q**.

....

....

(ii) Reflect shape **P** in the line $x = 1$

(5)

5.

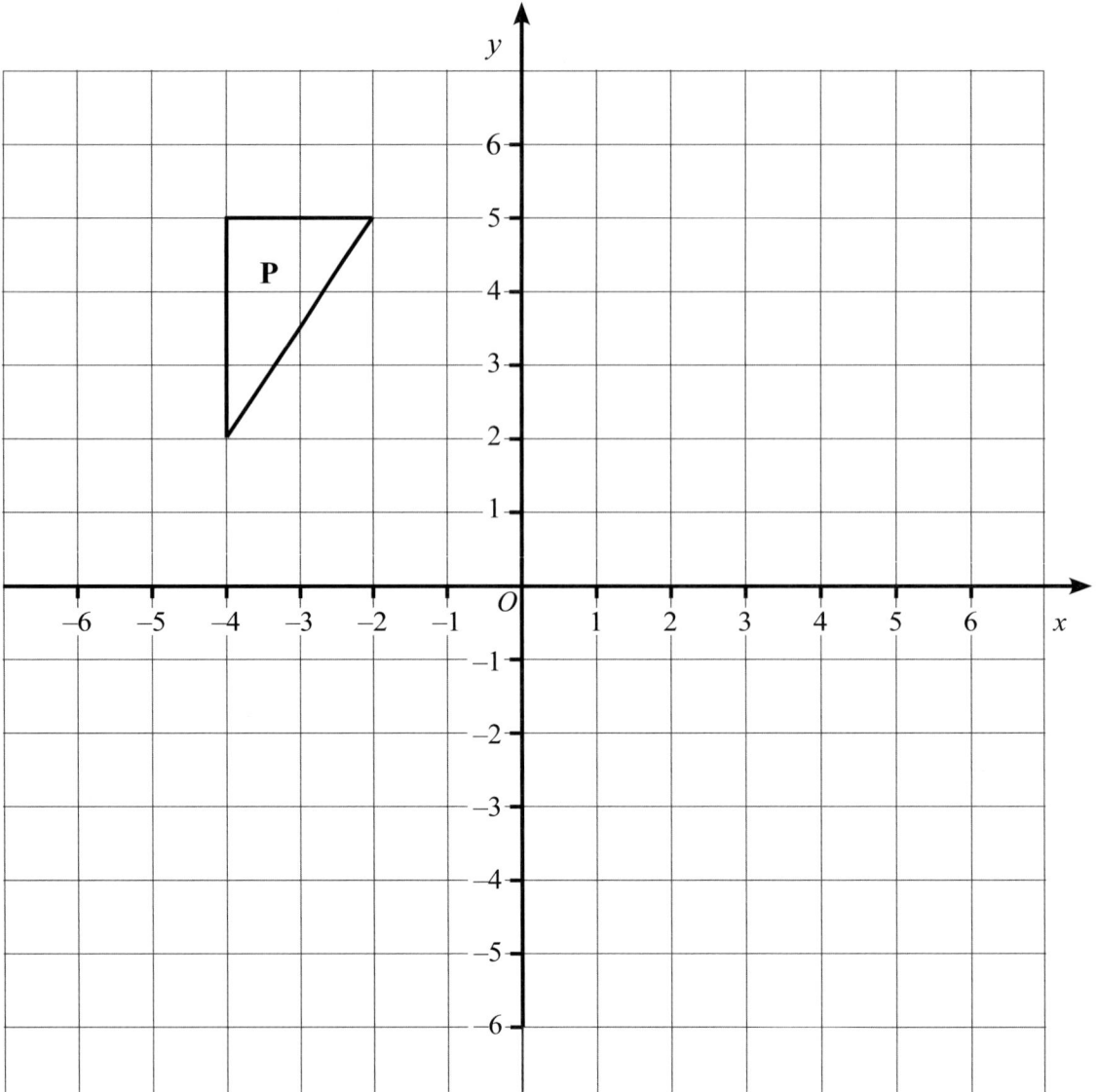

(a) Rotate triangle **P** 90° clockwise about the point (0, 2)
Label the new triangle **Q**.

(2)

(b) Translate triangle **P** by the vector $\begin{pmatrix} 5 \\ -6 \end{pmatrix}$

Label the new triangle **R**.

(1)

6. A shape has been drawn on a grid of centimetre squares.

On the grid, enlarge the shape with a scale factor of 2.

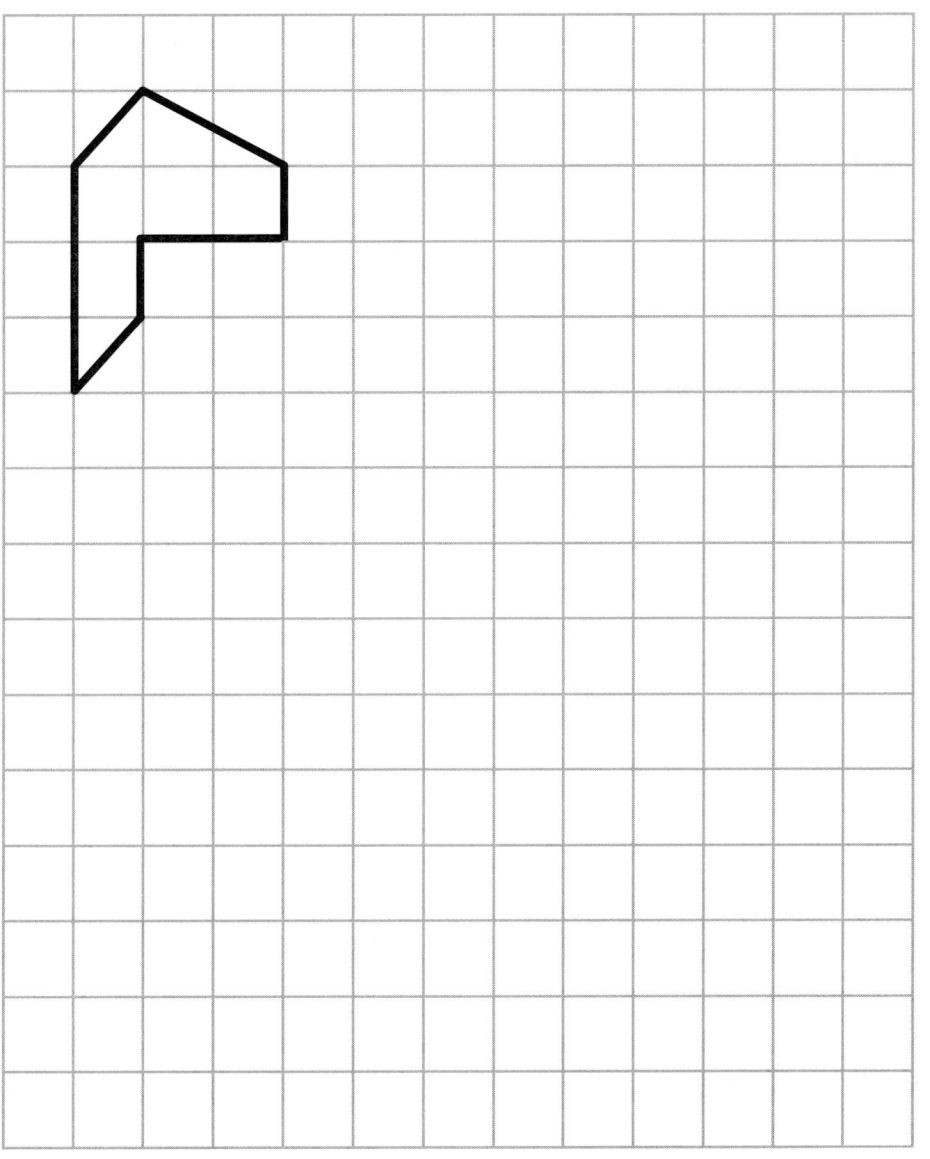

(2)

A03 - Shape, space and measures

7.

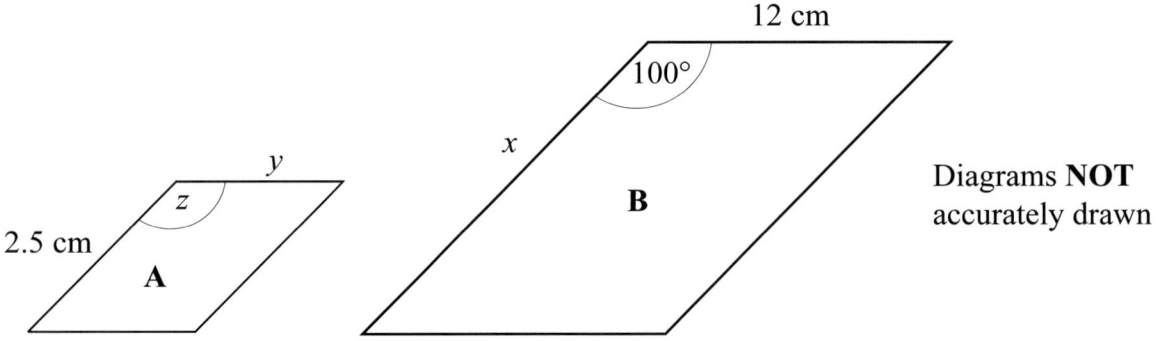

Diagrams **NOT** accurately drawn

Parallelogram **B** is an enlargement of parallelogram **A**.
The scale factor of the enlargement is 3.

(a) Find the length of the side marked x.

$x =$ .... cm

(1)

(b) Find the length of the side marked y.

$y =$ ....cm

(1)

(c) Find the size of the angle marked z.

$z =$ ....$°$

(1)

8.

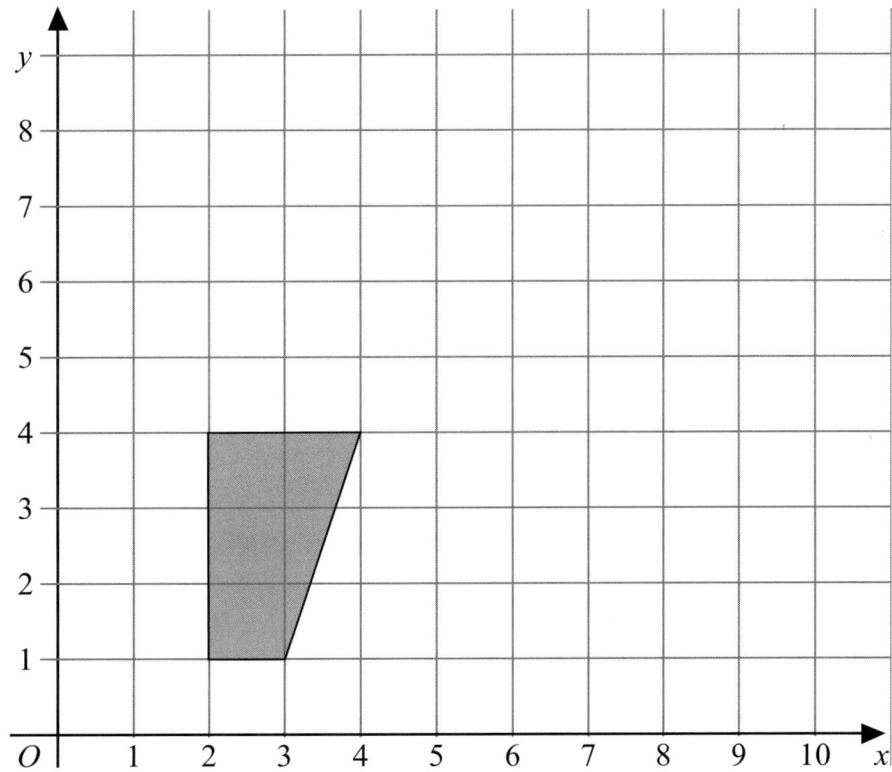

On the grid, enlarge the shaded shape by scale factor of 2, centre (1,1).

(3)

9.

The trapezium **T** is enlarged.

The line *PQ* becomes the line *XY*.

On the grid, complete the enlargement of trapezium **T**.

(2)

6 Measures and construction

❏ Measure a bearing

❏ Plot a point given the bearing and distance

❏ Calculate a bearing

❏ Draw two bearings to locate a 3rd on a scaled diagram

1.

(a) Measure and write down the bearing of Q from P.

....$^{\circ}$

(1)

(b) Find the bearing of T from R.

....$^{\circ}$

(1)

2. In the diagram, Point A marks the position of Prestwich.
The position of Radcliffe is to be marked on the diagram as *Point B*.

On the diagram, mark with a cross (×) the position of *B*, given that
    B is on a bearing of 320° from A and
    B is 6 cm from *A*.

N

×
A

(2)

3.   A lighthouse, *L*, is 3.2 km due West of a port, *P*.
A ship, *S*, is 1.9 km due North of the lighthouse, *L*.

Diagram **NOT**
accurately drawn

Find the bearing of the port, *P*, from the ship, *S*.
Give your answer correct to 3 significant figures.

.... °

(1)

4.

Diagram **NOT**
accurately drawn

Work out the bearing of

(i)   *B* from *P*,

.... °

(ii) *P* from *A*.

.... °

(3)

5.   *A* and *B* are two schools
B is due East of *A*.
C is another school.
The bearing of *C* from *A* is 064°.
The bearing of *C* from *B* is 312°.

Complete the scale drawing below.
Mark, with a cross (×), the position of the school *C*.

(2)

A03 - Shape, space and measures

6B - Scale and speed   *LIN 16, 24 MOD 17, 33*

- ❏ Measure or draw a scaled line
- ❏ Use a map and scale to measure and find distance
- ❏ Use simple scales in calculation
- ❏ Make simple scale drawings
- ❏ Estimate lengths in real life situations
- ❏ Identify unit to be used for a particular measurement
- ❏ Convert between metric units
- ❏ Use Speed = Distance ÷ Time in calculations

1.

P ———————————————————— Q

Measure the length PQ.

.... cm

(1)

2.

× York

× Leeds

Scale: 1 cm represents 3 km

Find the actual distance between Leeds and York.
Give your answer in kilometres.

.... km

(3)

3.

The scale of a map is 1 : 50 000

Work out the real distance 6 cm represents.
Give your answer in kilometres.

.....................km

(3)

4.

A map is drawn to a scale of 1:25 000

Two schools A and B are 12 centimetres apart on the map.

Work out the actual distance from A to B.
Give your answer in kilometres.

.... km

(3)

5.

The scale diagram shows a man and a dinosaur.

The man is 6 feet tall.

Estimate the height of the dinosaur:

(i) in feet,

.... feet

(ii) in metres.

.... metres

(2)

6.

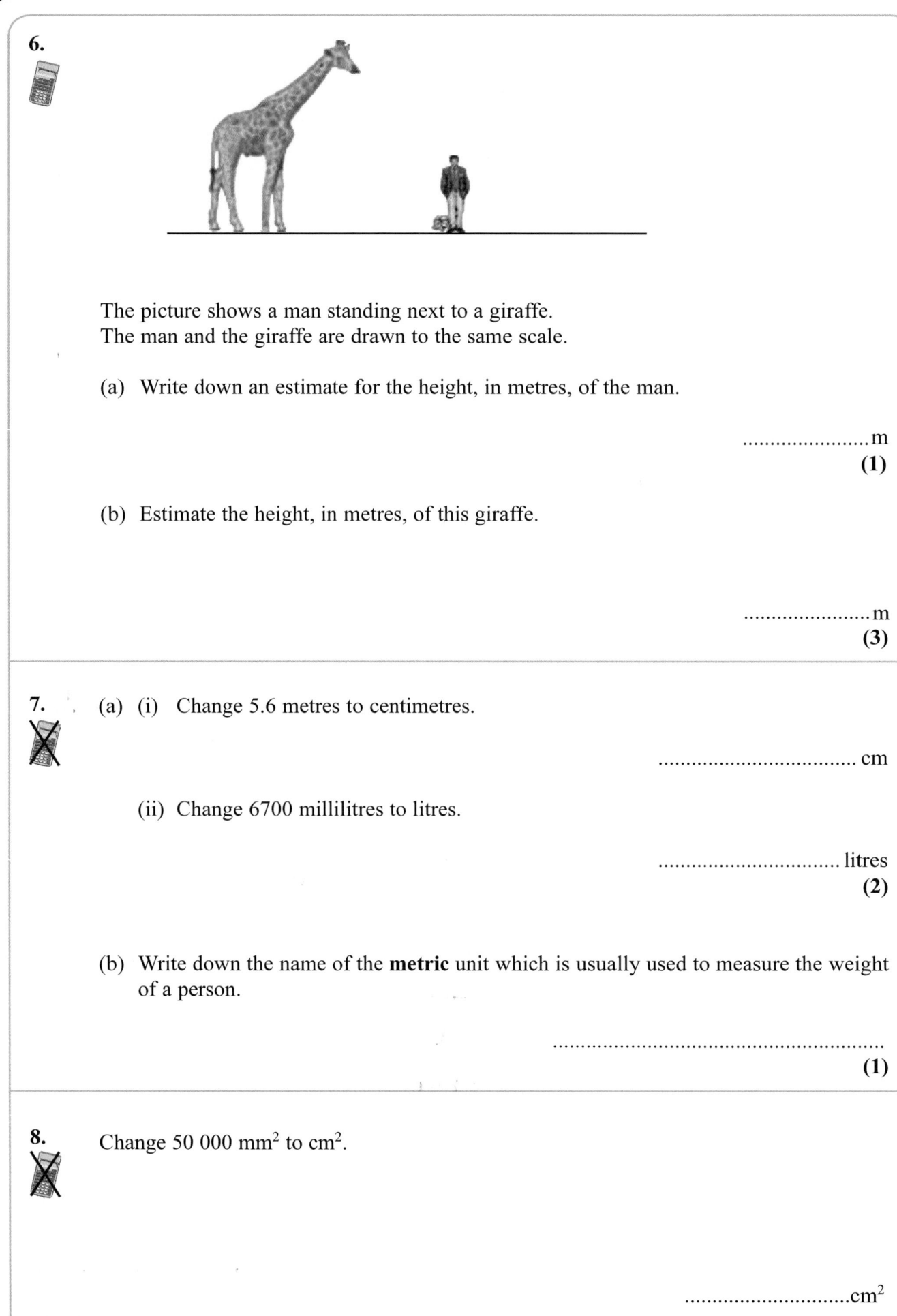

The picture shows a man standing next to a giraffe.
The man and the giraffe are drawn to the same scale.

(a) Write down an estimate for the height, in metres, of the man.

........................m

(1)

(b) Estimate the height, in metres, of this giraffe.

........................m

(3)

7. (a) (i) Change 5.6 metres to centimetres.

.... cm

(ii) Change 6700 millilitres to litres.

.... litres

(2)

(b) Write down the name of the **metric** unit which is usually used to measure the weight of a person.

....

(1)

8. Change 50 000 mm² to cm².

.............................cm²

(2)

9. Change 7 m^2 to cm^2.

(2)

10. (a) Complete the table by writing a sensible metric unit on each dotted line. The first one has been done for you.

The distance from London to Birmingham	179 kilometres
The weight of a twenty pence coin	5
The height of the tallest living man	232
The volume of lemonade in a glass	250

(3)

(b) Change 5000 metres to kilometres.

........................km

(1)

11. Ann drives 210 km in 2 hours 40 minutes.

Work out Ann's average speed.

....

(4)

A03 - Shape, space and measures

6C - Construction and drawings

LIN 3, 5, 29 MOD 7, 23, 37

❏ Measure and draw a line

❏ Mark the midpoint of a line

❏ Measure and draw angles

❏ Make accurate drawings of complex shapes

❏ Draw a circle or arc

❏ Construct formal configurations using straight edge and compasses

❏ Construct an equilateral triangle triangles given the length of all three sides

❏ Use given information to draw/construct loci

1.

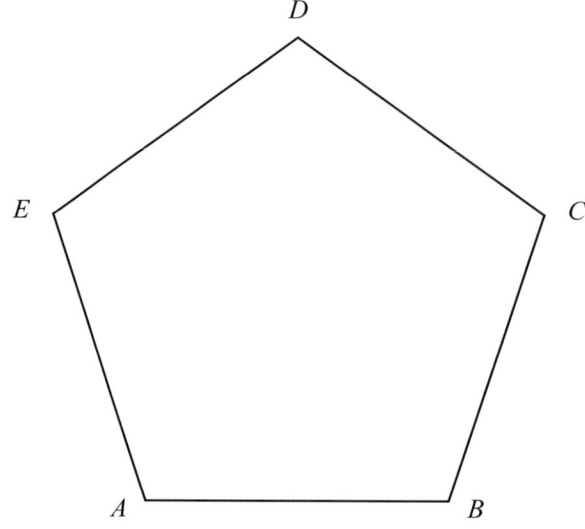

(a) (i) Measure the length of *AB*.

.... cm

(ii) Measure the size of angle *A*.

.... °

(2)

(b) In the space below, draw a line that is 12 cm long.

(1)

(c) Mark with a cross (✗) the midpoint of the line that you have drawn.

(1)

2.

Diagram **NOT**
accurately drawn

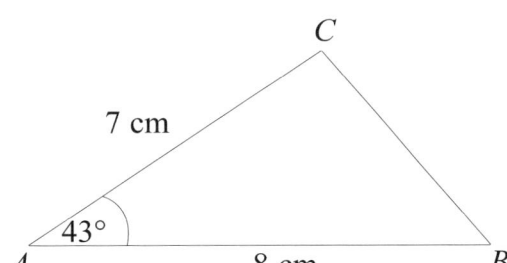

ABC is a triangle.
AB = 8 cm.
AC = 7 cm.
Angle *A* = 43°.

In the space below, make an accurate drawing of triangle *ABC*.
AB has been drawn for you.

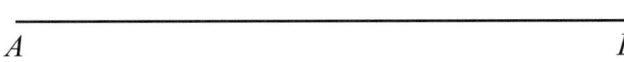

A   *B*

(2)

A03 - Shape, space and measures

3.

P ──────────────────────────────── Q

Use ruler and compasses to **construct** the perpendicular bisector of the line segment *PQ*.
You must show all your construction lines.

(2)

4. In the space below, use ruler and compasses to **construct** an equilateral triangle with
sides of length 6 centimetres.
You must show all construction lines.
One side of the triangle has already been drawn for you.

──────────────────────

(2)

5.

A

B
×

× C

Jill rolls a ball from point C.
At any point on its path, the ball is the same distance from point A and point B.

(a) On the diagram above, draw accurately the path that the ball will take.

(2)

(b) On the diagram, shade the region that contains all the points that are no more than 3 cm from point B.

(2)

6D - Area and perimeter *LIN 9 MOD 12*

- ❏ Estimate area by counting squares of simple shapes
- ❏ Find perimeter by counting squares
- ❏ Calculate the area and perimeter of a rectangle
- ❏ Use inverse rules of area formulae to find length of sides
- ❏ Calculate the area and perimeter of more complex shapes

- ❏ Find the area of a triangle using $A = \frac{1}{2}$ base x height
- ❏ Find the area and perimeter of compound shapes
- ❏ Find the area of a trapezium by substitution in the formula
- ❏ Convert between units of area

1. Here are 8 shapes on a grid of centimetre squares.

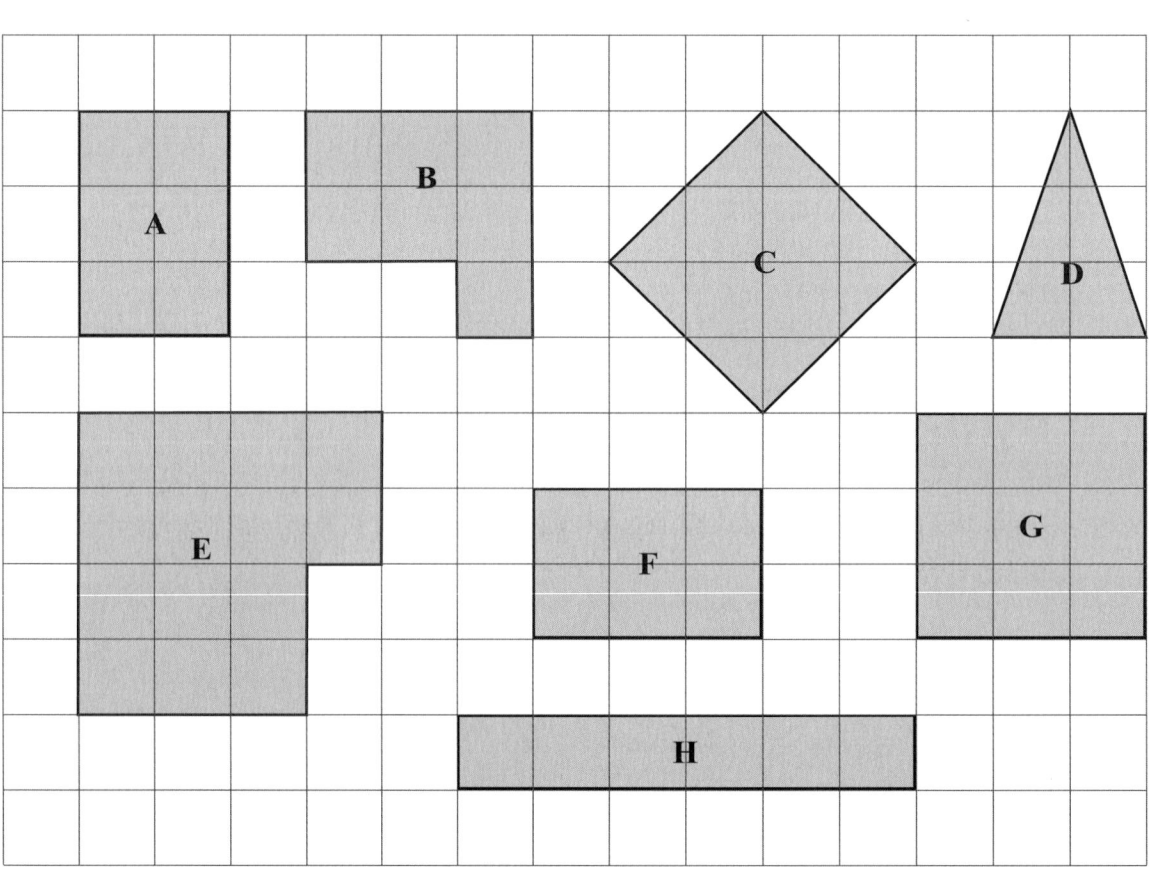

Find the area of shape **C**.

.... cm²

(2)

2.   A shaded shape has been drawn on the centimetre grid.

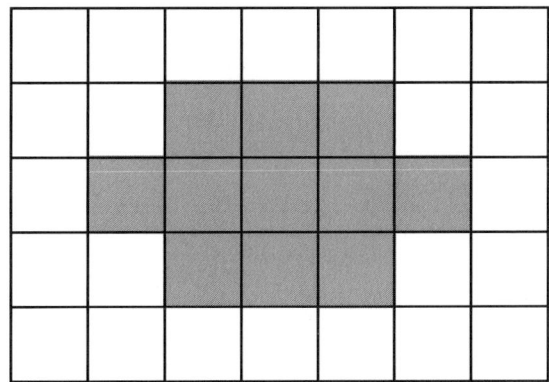

(i)   Find the area of the shaded shape.

.... cm^2

(ii) Find the perimeter of the shaded shape.

.... cm

(2)

3.   A shaded shape has been drawn on a grid of centimetre squares.

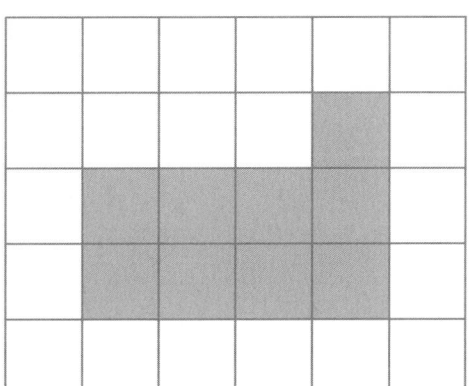

(a) Find the perimeter of the shaded shape.

.... cm

(1)

Another shaded shape has been drawn on a grid of centimetre squares.

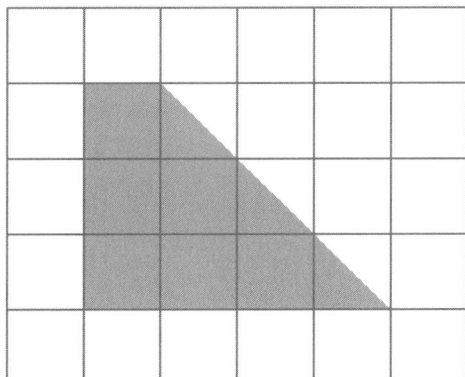

(b) Find the area of the shaded shape.

.... cm^2

(2)

A03 - Shape, space and measures

4.    A shape has been drawn on a grid of centimetre squares.

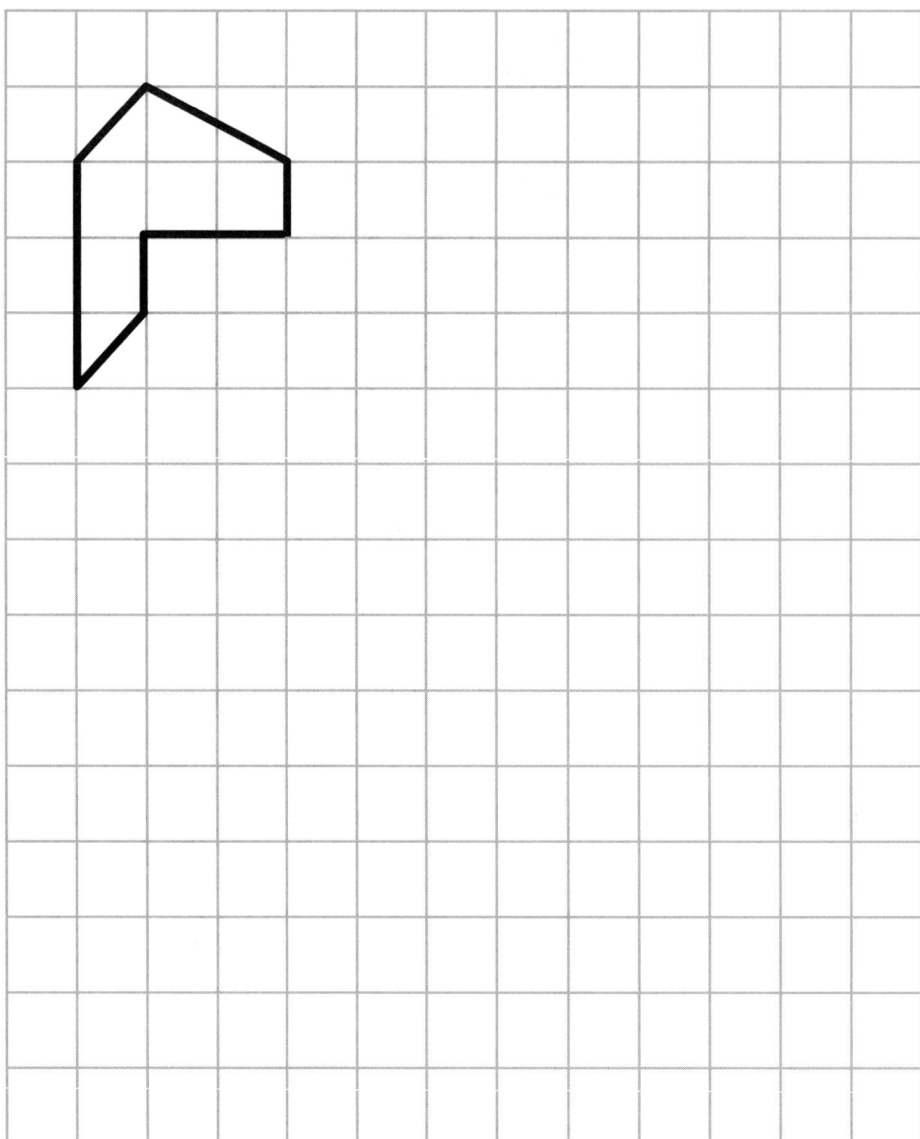

Work out the area of the shape.
State the units with your answer.

....

(3)

5.

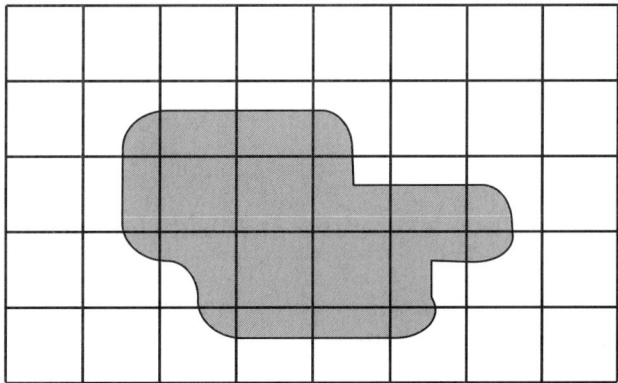

The shaded shape on the diagram represents the surface of a lake in winter.
The lake is drawn on a cm² grid.

(a) Estimate the area, in cm², of the shaded shape.

.... cm²

(2)

Each square on the grid represents a square with sides of length 100 m.

(b) Work out the area, in m², represented by one square on the grid.

.... m²

(2)

(c) Estimate the area, in m², of the lake.

.... m²

(1)

A03 - Shape, space and measures

6.   Here is a rectangle.

3 cm

Diagram **NOT** accurately drawn

4 cm

(a)   Work out the area of the rectangle.

.... cm²

(2)

(b)   Work out the perimeter of the rectangle.

.... cm

(1)

7. The diagram shows a rectangular field.

Diagram **NOT** accurately drawn

The length of the field is 54.5 m.
The width of the field is 35.5 m.

The field is for sale.
Mrs Fox wants to buy the field.
She also wants to plant a hedge along the perimeter.

The field costs £11.44 per square metre.
Each metre length of hedge costs £4.81

Mrs Fox has £23 000

Has Mrs Fox enough money to buy the field and plant the hedge?

You must show the working you use to make your decision.

(6)

A03 - Shape, space and measures

8. The diagram shows a Tangram.

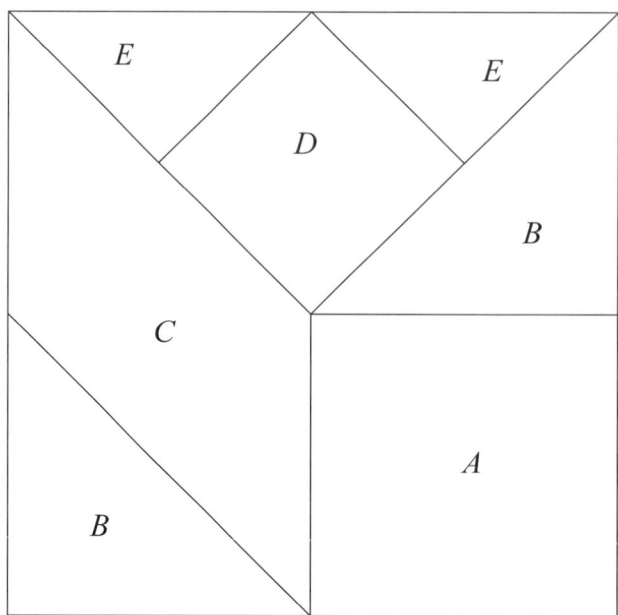

Diagram **accurately** drawn

The Tangram is a large square that is made up from

one square *A*,
two triangles *B*,
one parallelogram *C*,
another square *D* and
two small triangles *E*.

The total area of the Tangram is 64 cm².

Find the area of

   (i)   square *A*,

                                  .... cm²
                                                 (1)

   (ii) triangle *B*,

                                  .... cm²
                                                 (1)

   (iii) parallelogram *C*.

                                  .... cm²
                                                 (1)

9.

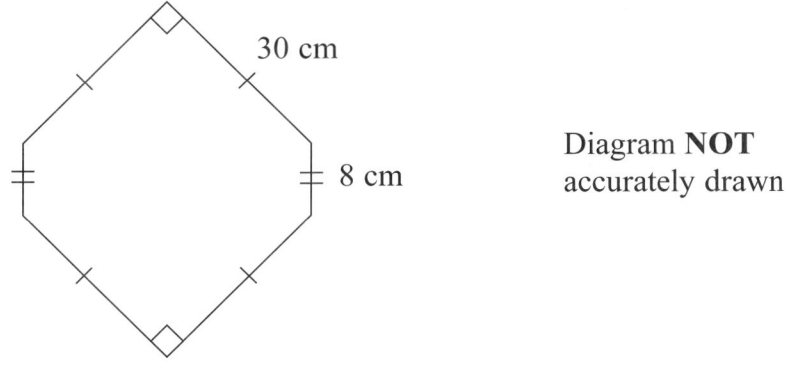

30 cm

8 cm

Diagram **NOT**
accurately drawn

The diagram shows the lengths of two of the sides of the shape.

Work out the perimeter of the shape.

.....................cm

(2)

10.

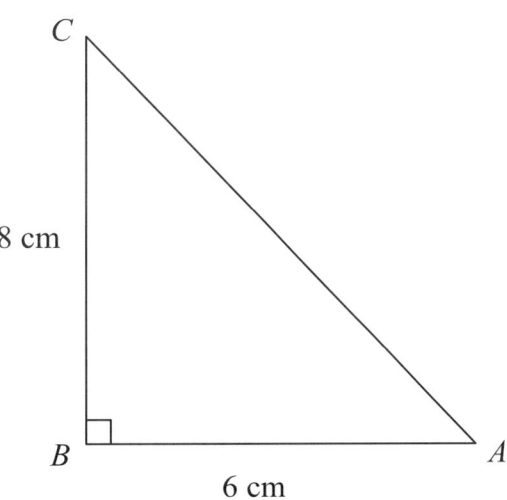

C

8 cm

B

6 cm

A

Diagram **NOT**
accurately drawn

Calculate the area of the triangle.

.... cm²

(2)

11. The diagram shows a 6-sided shape made from a rectangle and a right-angled triangle.

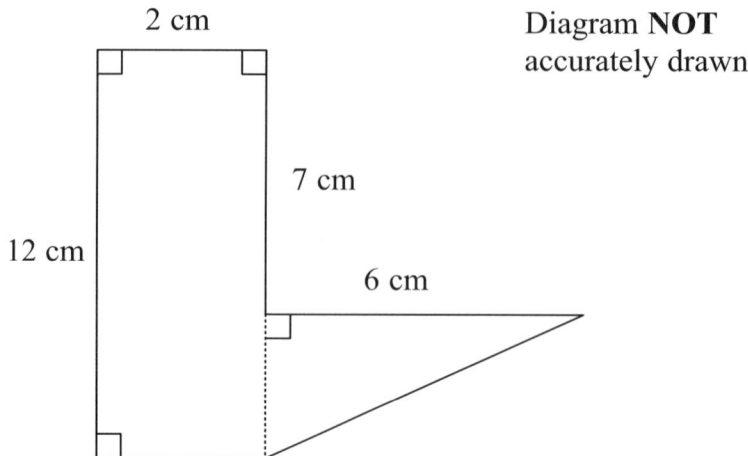

2 cm

Diagram **NOT**
accurately drawn

7 cm

12 cm

6 cm

Work out the total area of the 6-sided shape.

...............................cm²

(3)

12. The diagram shows a trapezium of height 3 m.

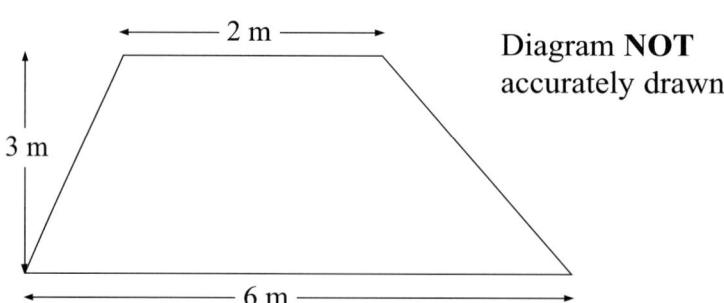

2 m

Diagram **NOT**
accurately drawn

3 m

6 m

Find the area of this trapezium.

State the units with your answer.

....

(3)

6E - Circle measure

LIN 22 MOD 31

❏ Use the formulae $C=2\pi r$ or $A=\pi r^2$ to find
   the circumference and area of circles

1.

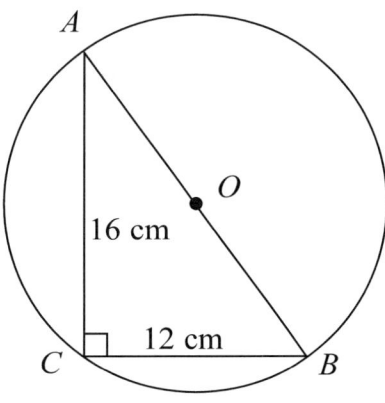

Diagram **NOT**
accurately drawn

The diagram shows triangle ABC and a circle, centre O.
A, B and C are points on the circumference of the circle.
AB is a diameter of the circle.

$AC = 16$ cm and $BC = 12$ cm.

Work out the area of the circle.
Give your answer correct to 3 significant figures.

.... cm^2

(3)

A03 - Shape, space and measures

4A - Angles

1. (a) isosceles
   (b) acute
   (c) obtuse

2. (a) (i) 48°
       (ii) alternate angles

   (b) (i) 30°
       (ii) corresponding angles

3. (a) (i) 103°
       (ii) corresponding angles

   (b) 26°

4. 137°
   Interior angles in a quadrilateral = 360° and angles on straight line = 180°

5. A right angle = 90°
   Sum of angles at a point is 360°

6. (i) 140°
   (ii) Angles on straight line = 180°

7. (a) (i) 60°
       (ii) Triangle is equilateral
   (b) 150°

8. 287°

9. (i) 100°
   (ii) Base angles of isosceles triangles are equal *or* Angles in a triangle add up to 180°

10. 94°

11. 22.5°

12. 60°

4B - Properties of shapes

1. (i)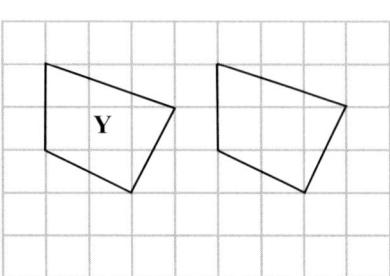
   (ii) right angle
   (iii) equilateral

2. (a) isosceles   (b) acute   (c) obtuse

3. (a) isosceles
   (b) A and F

4. (a) A and E
   (b)

   Y

5. hexagon

6. 1st - sector
   3rd - tangent
   4th - segment
   5th - diameter

7.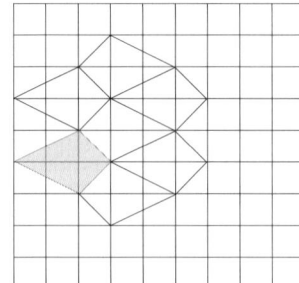

8.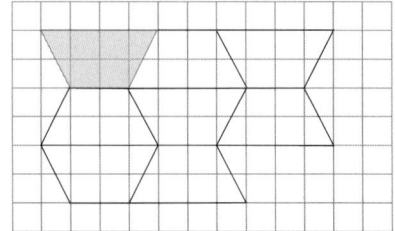

4C - 3-D Shapes and volumes

1. (i) Cylinder
    (ii) Cuboid

2. 1 to 5
    2 to 3
    3 to 4
    5 to 1

3.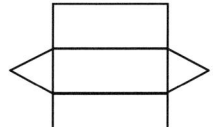

4.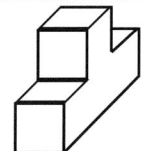

5. 264 cm^2

6. 12 cm^3

7. (i) 7   (ii) 15   (iii) 10

8. 1000

9. (a) 4 cm
    (b) 100

4D - Pythagoras

1. 10.9 cm

2. 20 cm

3. 13.7 cm

4. 10 cm

5A - Symmetry

1.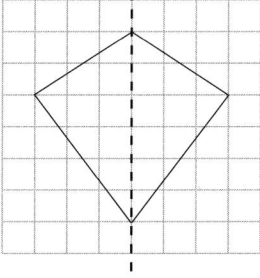

2.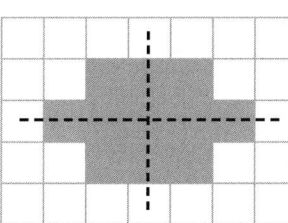

3. (i)
    (ii) right angle
    (iii) equilateral

4.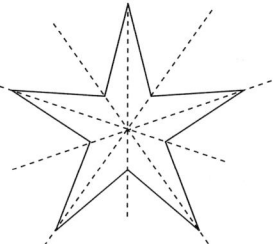

5. (a) 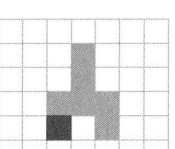   (b)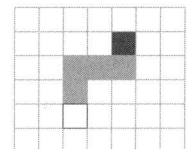

6. B

7. (i) 18   (ii) 11 (or 88)   (iii) 69

8.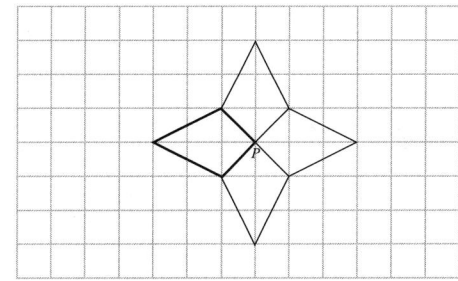

9. 5

5B - Coordinates

1. (a) (i) (0, 2)

     (ii) (4, 1)

(b)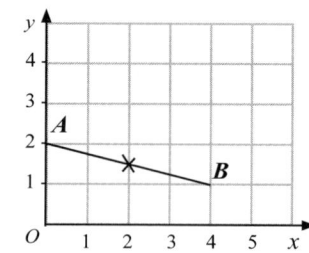

2. (1, 2)

5C - Transformations

1.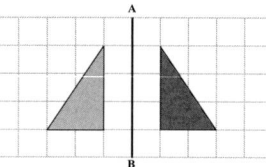

2.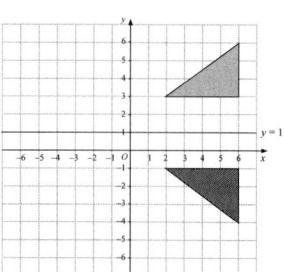

3. (a)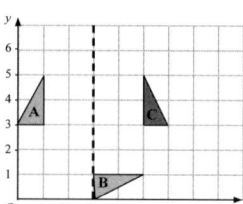

   (b) Reflection in the line $y = x$

4. (i) Enlargement by scale factor 2, centre (0,0)

   (ii)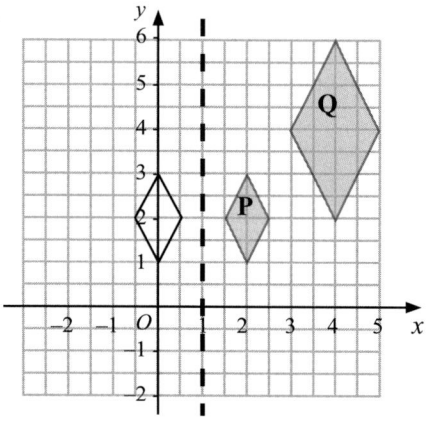

5. (a)

   (b)

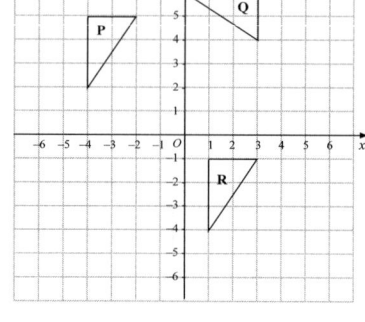

6.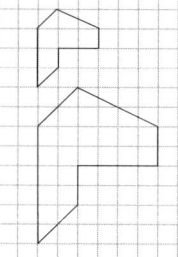

7. 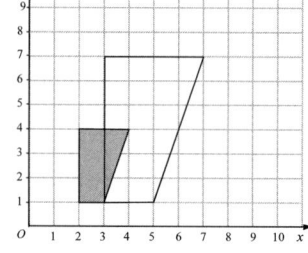   100°

8.

9.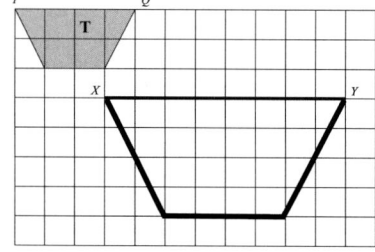

6A - Bearings

1. (a) 045°

   (b) 270°

2.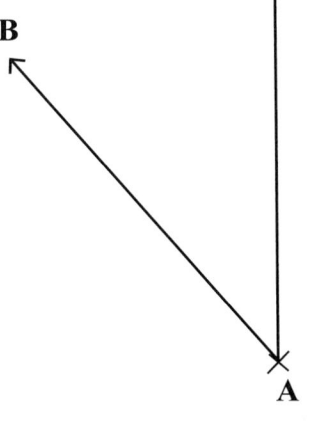

3. 121°

4. (i) 222°

   (ii) 243°

5.

6B - Scale and speed

1. 7.5 cm

2. 36 km

3. 3 km

4. 3 km

5. (i) 20 ft
   (ii) 6 m

6. (a) 1.5 - 2.0 m
   (b) 3 - 6 m

7. (a) (i) 560 cm   (ii) 6.7 litres
   (b) kg

8. 500 cm^2

9. 70 000 cm^2

10. (a) Grams, g
       centimetres, cm
       millilitres, ml, cm^3
    (b) 5

11. 78.75 km/h

6C - Construction and drawings

1. (a) (i) 4 cm ± 0.2

        (ii) 108° ±2

    (b) & (c)

2.

4.

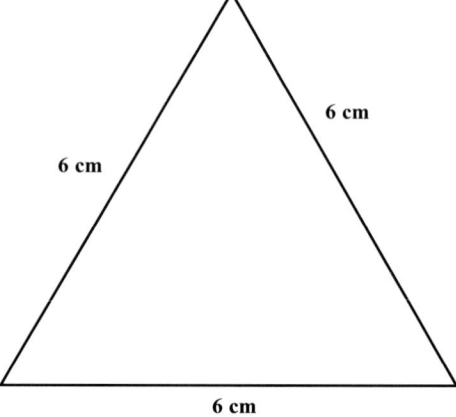

5.

3.

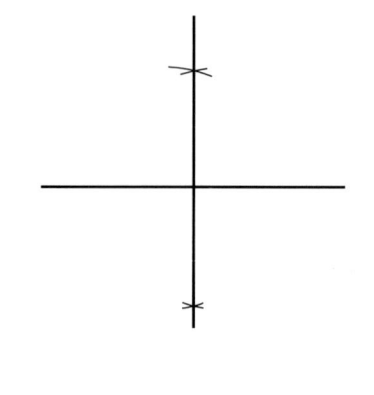

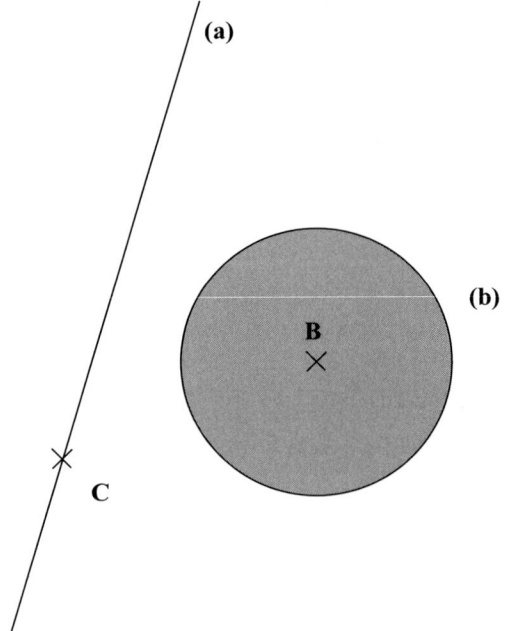

6D - Area and perimeter

1. 8 cm^2

2. (i) 11 cm^2
    (ii) 16 cm

3. (a) 14 cm
    (b) 7.5 cm^2

4. 6 cm^2

5. (a) 11 cm^2
    (b) $10\ 000 \text{ m}^2$
    (c) $110\ 000 \text{ m}^2$

6. (a) 12 cm^2
    (b) 14 cm

7. Total cost: £22 999.34
    Yes, Mrs Fox can buy both the field and the
    hedge

8. (i) 16 cm^2
    (ii) 8 cm^2
    (iii) 16 cm^2

9. 136 cm

10. 24 cm^2

11. 39 cm^2

12. 12 m^2

6E - Circle measure

1. 314 cm^2

A04
Handling data

Contents

7 Specifying the problem and planning

7A - Surveys

❑ Design questions for a questionnaire with a response section

❑ Explain deficiencies in questions

1. Janie wants to collect information about the amount of sleep the students in her class get.

   Design a suitable question she could use.

   (2)

2. A driving test centre is designing a questionnaire to find out how many hours of driving lessons people have before they take the test.

   They have designed this question.

   "How long have you been having driving lessons?"

   (a) Write down one thing that is wrong with this question.

   ....

   ....

   (1)

   (b) Design a better question for the driving centre to use.
   You should include some response boxes.

   (2)

3. Petros wants to find out how teenagers communicate with each other.
He designs a questionnaire.
Here are two of his questions.
The questions are **not** suitable.
For each question, write down a reason why.

(i) Do you prefer to communicate with your best friend by mobile phone or by e-mail?

Yes ☐   No ☐

Reason ....

....

(ii) How many e-mail addresses do you have?

1 ☐   2 ☐   3 ☐   4 ☐

Reason ....

....

(2)

4. Daniel is conducting a survey into the amount of money that teenagers spend on magazines.

He uses this question on a questionnaire.

"How much money do you spend on magazines?"

£1   £2   £3

☐   ☐   ☐

Write down **two** things that are wrong with this question.

....

....

....

(2)

8 Collecting data

8A - Collecting data   LIN 2, 18, 23 MOD 1, 3, 4

- ❏ Collect discrete data in a tally chart
- ❏ Find total frequency from a tally chart
- ❏ Interpret data from a grouped frequency table
- ❏ Construct a data collection sheet
- ❏ Complete a 2-way table from information given
- ❏ Use and extract information from lists and tables

1. Daniel carried out a survey of his friends' favourite flavour of crisps.

Here are his results.

Plain	Chicken	Bovril	Salt & Vinegar	Plain
Salt & Vinegar	Plain	Chicken	Plain	Bovril
Plain	Chicken	Bovril	Salt & Vinegar	Bovril
Bovril	Plain	Plain	Salt & Vinegar	Plain

(a) Complete the table to show Daniel's results.

Flavour of crisps	Tally	Frequency
Plain		
Chicken		
Bovril		
Salt & Vinegar		

(3)

(b) Write down the number of Daniel's friends whose favourite flavour was Salt & Vinegar.

....

(1)

(c) Which was the favourite flavour of most of Daniel's friends?

....

(1)

2.  Luigi and Francesca carried out a survey of the vehicles passing their house. Here are their results.

Car	Van	Lorry	Bike	Bus	Car
Van	Car	Car	Van	Lorry	Bike
Bike	Bike	Van	Lorry	Bike	Car
Car	Bus	Lorry	Car	Lorry	Bike

Complete the tally column and frequency column in the frequency table.

Type of vehicle	Tally	Frequency
Car		
Van		
Lorry		
Bike		
Bus		

(2)

3. The table gives information about the number of goals scored by a football team in each match during a season.

Number of goals	Number of matches
0	9
1	8
2	12
3	5

Work out the total number of goals scored by the football team during the season.

....

(2)

4. Amberish is going to carry out a survey about zoo animals.

He decides to ask some people whether they prefer
lions, tigers, elephants, monkeys or giraffes.

Design a data collection sheet that he can use to carry out his survey.

(3)

5. 60 British students each visited one foreign country last week.

This two-way table shows information about which countries the students visited.

	France	Germany	Spain	Total
Female			9	34
Male	15			
Total		25	18	60

Complete the two-way table.

(3)

6. Some bulbs were planted in October.
The ticks in the table shows the months in which each type of bulb grows into flowers.

		Month					
		Jan	Feb	March	April	May	June
Type of bulb	Allium					✓	✓
	Crocus	✓	✓				
	Daffodil		✓	✓	✓		
	Iris	✓	✓				
	Tulip				✓	✓	

(a) In which months do tulips flower?

....

(1)

(b) Which type of bulb flowers in March?

....

(1)

(c) In which month do most types of bulb flower?

....

(1)

(d) Which type of bulb flowers in the same months as the iris?

....

(1)

7.

The chart shows the shortest distances, in kilometres, between pairs of cities.
For example, the shortest distance between London and Manchester is 290 km.

London				
196	**Nottingham**			
290	101	**Manchester**		
325	158	56	**Liverpool**	
639	446	346	348	**Glasgow**

(a) Write down the shortest distance between **Nottingham** and **Liverpool**.

.... km

(1)

Daniel drives from London to Manchester by the shortest route.
He drives 137 km and stops for a rest.

(b) Work out how many more kilometres he must drive.

.... km

(2)

(c) Write down the names of the two cities which are the **least** distance apart.

.... and ....

(1)

9 Processing and representing data; Interpreting and discussing results

9A - Representing and interpreting data   *LIN 14, 18, 30 MOD 2, 3, 5*

❏ Draw pictograms

❏ Interpret pictograms

❏ Draw bar charts

❏ Interpret and compare bar charts, including dual bar charts

❏ Construct a pie chart from given data

❏ Use a pie chart

❏ Interpret a stem and leaf diagram to find median or range

❏ Construct a stem and leaf diagram

❏ Plot points and drawing and interpreting a scatter graph

❏ Find correlation from a scatter graph

❏ Draw and use lines of best fit to find and predict missing values

---

1. The pictogram shows the number of golfers who played at a golf club last week on Saturday, Sunday and Monday.

Saturday ⊕ ⊕ ⊕

Sunday ⊕ ⊕ ⊕

Monday ⊕ ⊕ ⊖

Tuesday

Key
⊕ Represents 20 golfers

(a) How many golfers played on Sunday?

....
(1)

(b) How many golfers played on Monday?

....
(1)

On Tuesday, 35 golfers played.

(c) Complete the pictogram.

(1)

2.

Jerry recorded the colour of each of the cars he saw one morning.

The bar chart shows this information.

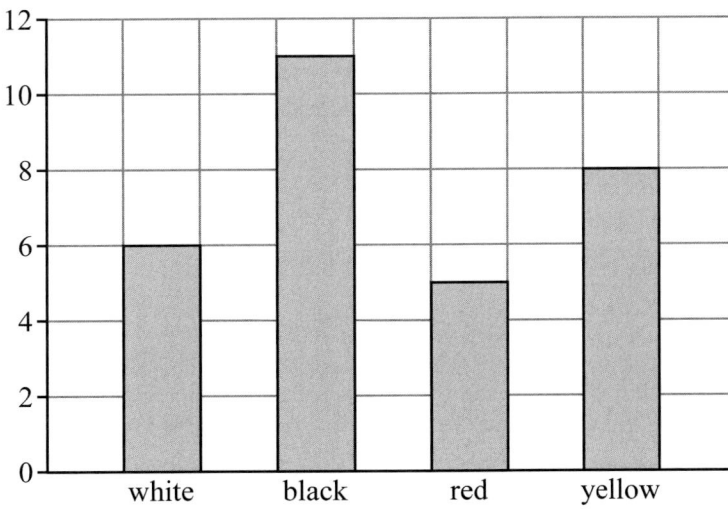

(a) Write down the number of red cars.

....

(1)

(b) Which was the most popular colour of car?

....

(1)

3. The table gives information about the medals won by Austria in the 2002 Winter Olympic Games.

Medal	Frequency	
Gold	3	
Silver	4	
Bronze	11	

Draw an accurate pie chart to show this information.

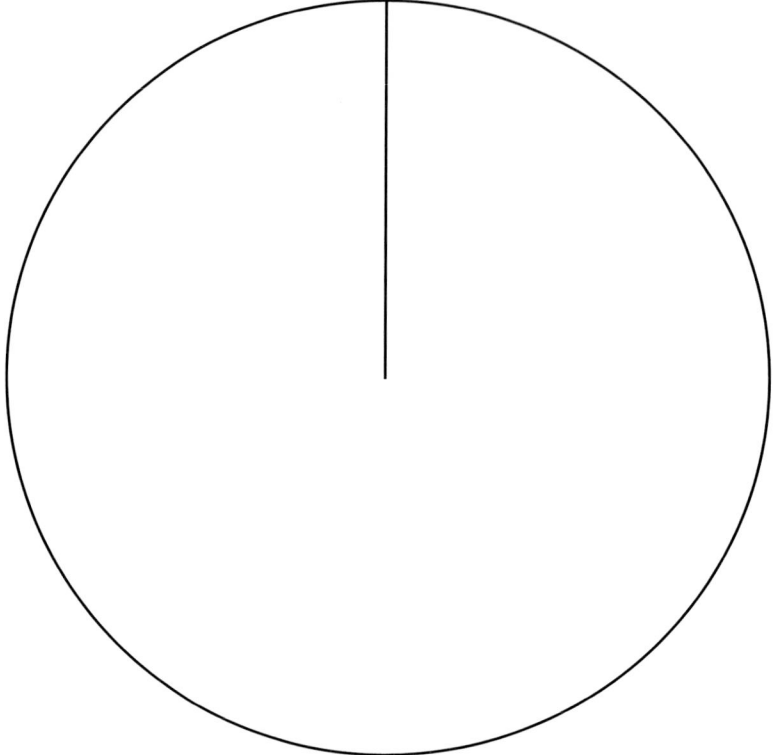

(4)

4. Sandra carries out a survey of 90 Year 11 students.
She asks them their favourite snack.

She draws this accurate pie chart.

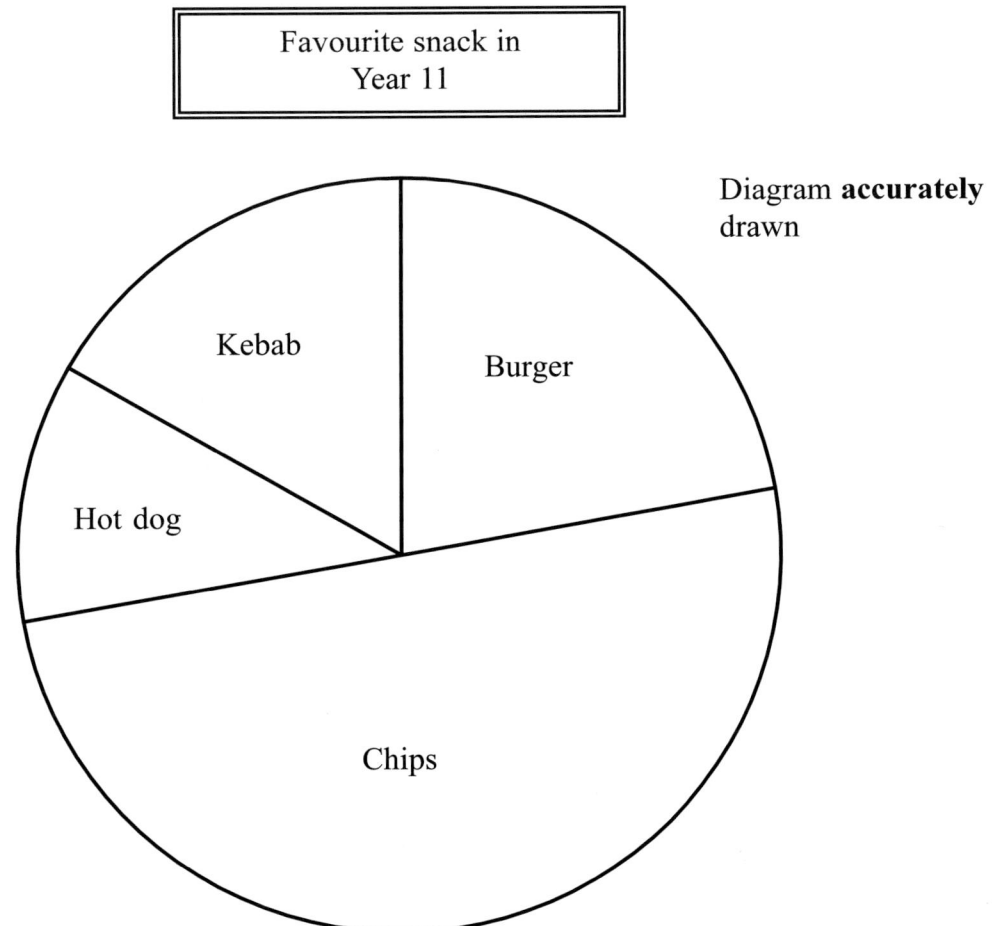

Favourite snack in
Year 11

Diagram **accurately**
drawn

Use the pie chart to complete the table.

Favourite snack in Year 11	Frequency	Angle
Burger	20	
Chips	45	180°
Hot dog		
Kebab		
Total	90	

(4)

5. A shop sells cookers.
The pie chart shows some information about the number of cookers the shop sold in one year.

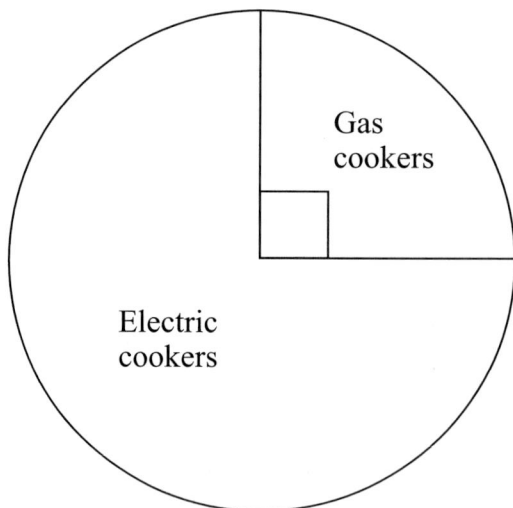

The shop sold 150 gas cookers.
Work out the total number of cookers the shop sold.

....

(2)

6. The stem and leaf diagram shows information about the pulse rate of each of 15 students.

Pulse rate

```
5 | 6 8
6 | 0 2 3 8
7 | 1 4 6 6 8
8 | 7 8 9
9 | 7
```

Key:
5	6 means 56

Work out the range of the pulse rates.

....

(1)

7. Tom collects information about the age and cost of some Ford Mondeo cars.

He plots a scatter graph of his results. Here is his graph.

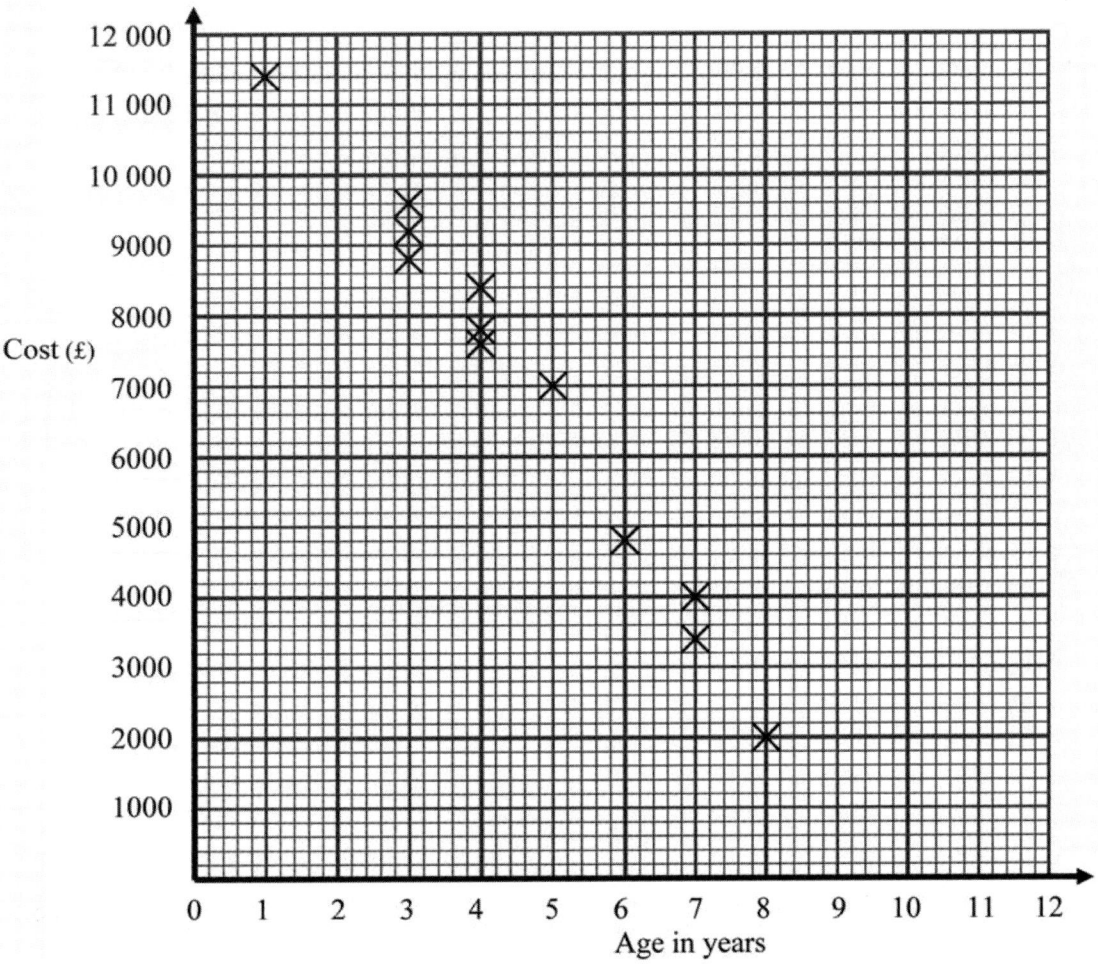

Tom collects data on 3 more Ford Mondeo cars.

Age	2	7	9
Cost (£)	10 000	3000	1000

(a) What type of correlation does Tom's scatter graph show?

....

(1)

(b) Draw in a line of best fit onto the scatter graph.

(1)

A04 - Handling data

9B - Averages

❏ Find the mode

❏ Find the modal class from a grouped frequency table

❏ Find the median

❏ Find the mean

❏ Find an estimate for the mean from a grouped frequency table

❏ Find the range

1. Chloe made a list of her homework marks.

$$4 \quad 5 \quad 5 \quad 5 \quad 4 \quad 3 \quad 2 \quad 1 \quad 4 \quad 5$$

(a) Write down the mode of her homework marks.

....
(1)

(b) Work out her mean homework mark.

....
(2)

2. The table gives information about the ages of 160 employees of an IT company.

Age (A) in years	Frequency
$15 < A \leqslant 25$	44
$25 < A \leqslant 35$	56
$35 < A \leqslant 45$	34
$45 < A \leqslant 55$	19
$55 < A \leqslant 65$	7

Write down the modal class interval.

....
(1)

3. Tom recorded the shoe size of five of his friends.
Here are his results.

8   9   3   4   7

(a) Work out the median shoe size.

....

(2)

Another friend has a shoe size of 8

(b) Work out the median shoe size of all **six** friends of Tom.

....

(2)

4. Emma repairs bicycles.
She keeps records of the cost of the repairs.

The table gives information about the costs of all repairs which she carried out in one week.

Cost (£C)	Frequency
$0 < C \leqslant 10$	3
$10 < C \leqslant 20$	7
$20 < C \leqslant 30$	6
$30 < C \leqslant 40$	8
$40 < C \leqslant 50$	9

Find the class interval in which the median lies.

....

(2)

A04 - Handling data

5. Fred did a survey on the areas of pictures in a newspaper.
The table gives information about the areas.

Area (A cm^2)	Frequency
$0 < A \leqslant 10$	38
$10 < A \leqslant 25$	36
$25 < A \leqslant 40$	30
$40 < A \leqslant 60$	46

Work out an estimate for the mean area of a picture.

.... cm^2

(4)

6. Here is a list of the numbers of golfers who played at the club each day over a two week period.

63   72   42   51   38   56   45   67   82   45   64   77   56   49

(i) Find the median.

....

(ii) Find the range.

....

(3)

7. The table shows information about the number of hours that 120 children used a computer last week.

Number of hours (h)	Frequency
$0 < h \leqslant 2$	10
$2 < h \leqslant 4$	15
$4 < h \leqslant 6$	30
$6 < h \leqslant 8$	35
$8 < h \leqslant 10$	25
$10 < h \leqslant 12$	5

Work out an estimate for the mean number of hours that the children used a computer. Give your answer correct to 2 decimal places.

.... hours

(4)

8. The table gives information about the time taken by 20 students to travel to school.

Time (t minutes)	Frequency	
$0 < t \leqslant 5$	2	
$5 < t \leqslant 10$	8	
$10 < t \leqslant 15$	4	
$15 < t \leqslant 20$	3	
$20 < t \leqslant 25$	3	

Work out an estimate for the mean time.

.... minutes

(4)

A04 - Handling data

❏ Find the least likely outcome of an event

❏ List all outcomes of an event and form a sample space

❏ Use probability scales

❏ Make estimates of probability from a pie chart, diagram, or table

❏ State probability of an outcome of an event happening

❏ Use the fact that the sum of probabilities is always 1

❏ Find probability from a 2-way table

❏ Find probability using 1 / n

❏ Find probability using a / n

❏ Select a prime numbers from a list

❏ Use relative frequency to estimate probability

1.

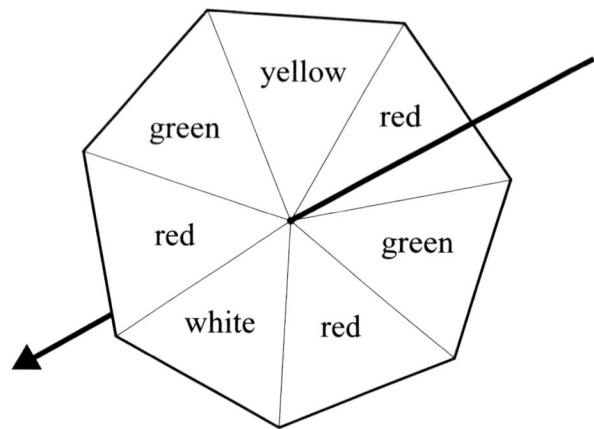

Here is a fair 7-sided spinner.
The spinner is to be spun once.
The spinner will land on one of the colours.

(a) On which colour is the spinner most likely to land?

....

(1)

(b) Write down the probability that the spinner will land on green.

....

(1)

2. There are three beads in a bag.
One bead is red, one bead is white and one bead is yellow.

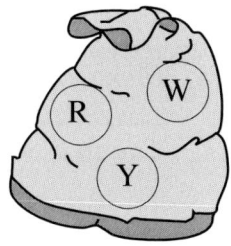

Sarah takes, at random, a bead from the bag.
She looks at its colour and then puts the bead back in the bag.

(a) On the probability line,

   (i) mark with the letter R the probability that Sarah takes a red bead.

   (ii) mark with the letter B the probability that Sarah takes a black bead.

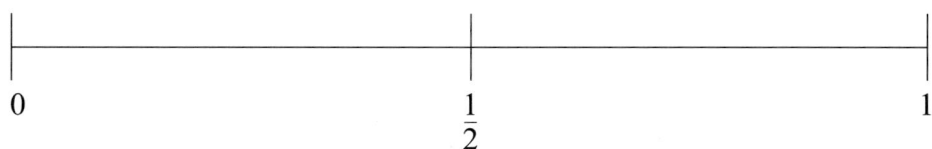

  0                               $\frac{1}{2}$ 1

(2)

There are also three beads in a box.
One bead is green, one bead is pink and one bead is blue.

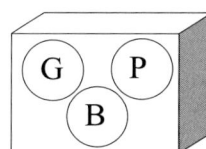

Without looking, Saskia takes, at random, one bead from the bag
and one bead from the box.
One possible outcome for the two beads she takes is (red, green).

(b) List all the possible outcomes.
One has already been done for you.

  (red, green) ....

  ....

  ....

(2)

A04 - Handling data

3. Phil rolls a dice and flips a coin.

(a) Make a list of all the possible combinations he could get.

The first one has been done for you.

(6, head) ....

....

....

(2)

Phil rolls a dice and flips a coin once.

(b) Work out the probability that he gets a 6 and a head.

....

(1)

4. Some bulbs were planted in October.
The ticks in the table shows the months in which each type of bulb grows into flowers.

		Month					
		Jan	Feb	March	April	May	June
Type of bulb	Allium					✓	✓
	Crocus	✓	✓				
	Daffodil		✓	✓	✓		
	Iris	✓	✓				
	Tulip				✓	✓	

Ben puts one of each type of these bulbs in a bag.
He takes a bulb from the bag without looking.

(i) Write down the probability that he will take a crocus bulb.

....

(ii) On the probability scale, mark with a cross (×) the probability that he will take a bulb which flowers in February.

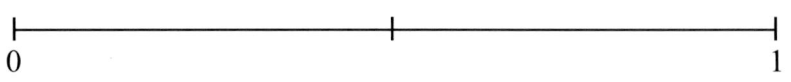

0   1

(2)

5. 60 British students each visited one foreign country last week.

This two-way table shows information about which countries the students visited.

	France	Germany	Spain	Total
Female	2	23	9	34
Male	15	2	9	26
Total	17	25	18	60

One of the students is picked at random.

Write down the probability that this student visited Germany last week.

....

(1)

6. A box contains 50 counters.
There are 23 white counters, 19 black counters and 8 yellow counters.

Piero takes at random a single counter from the box.

Work out the probability that he takes a white counter or a yellow counter.

....

(2)

A04 - Handling data

7. Tony throws a biased dice 100 times.
The table shows his results.

Score	Frequency
1	12
2	13
3	17
4	10
5	18
6	30

He throws the dice once more.

(a) Find an estimate for the probability that he will get a 6

....

(1)

Emma has a biased coin.
The probability that the biased coin will land on a head is 0.7
Emma is going to throw the coin 250 times.

(b) Work out an estimate for the number of times the coin will land on a head.

....

(2)

8.

Here is a 4-sided spinner.

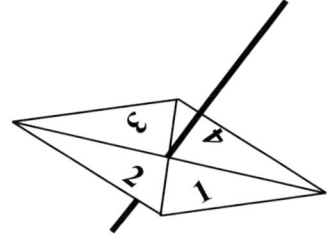

The sides of the spinner are labelled 1, 2, 3 and 4.
The spinner is biased.
The probability that the spinner will land on each of the numbers 2 and 3 is given in the table.
The probability that the spinner will land on 1 is equal to the probability that it will land on 4.

Number	1	2	3	4
Probability	x	0.3	0.2	x

(a) Work out the value of x.

$x =$ ....

(2)

Sarah is going to spin the spinner 200 times.

(b) Work out an estimate for the number of times it will land on 2.

....

(2)

A04 - Handling data

7A - Surveys

1. (a) A suitable question would be: How many hours of sleep do you get each night?

   You should include response boxes eg.

   Less than 6 hours   ☐

   6 - 8 hours   ☐

   More than 6 hours   ☐

2. (a) People would not give the number of hours, they would give the period of time during which they took lessons

   (b) A better question would be: How many hours have you spent on driving lessons? You should include response boxes e.g.

Less than 5 hours   ☐

5 hours but less than 10 hours   ☐

10 hours but less than 15 hours   ☐

15 or more hours   ☐

3. (i) The responses are not clear and the question should mention other ways of communicating

   (ii) The response boxes don't allow for all possible answers.

4. The question doesn't mention any particular period of time. The response boxes don't cover all possible responses.

8A - Collecting data

1. (a) Plain   8

   Chicken   3

   Bovril   5

   S & Vin   4

   (b) 4

   (c) Plain

2. Car   7

   Van   4

   Lorry   5

   Bike   6

   Bus   2

3. 47

4.

Animal	Tally	Total
Lions		
Tigers		
Elephants		
Monkeys		
Giraffes		

5.

	France	Germany	Spain	Total
Female	2	23	9	34
Male	15	2	9	26
Total	17	25	18	60

6. (a) April and May

   (b) Daffodil

   (c) Feb

   (d) Crocus

7. (a) 158 km

   (b) 153 km

   (c) Manchester and Liverpool

9A - Representing and interpreting data

1. (a) 60

   (b) 50

   (c)

2. (a) 5

   (b) black

3.

   (Pie chart: GOLD 60°, SILVER 80°, BRONZE 220°)

4.

Frequency	Angle
20	**80**
45	180
10	**40**
15	**60**
90	

5. 600

6. 41

7. (a) Negative

   (b)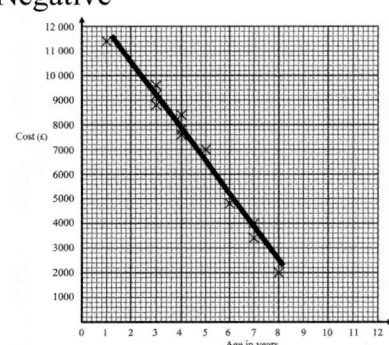

9B - Averages

1. (a) 5   (b) 3.8

2. 25 < A   35

3. (a) 7   (b) 7.5

4. 30 < C   40

5. 27.3 cm^2

6. (i) 56   (ii) 44

7. 6.08 hours

8. 11.75 mins

9C - Probability

1. (a) red

   (b) $\frac{2}{7}$

2. (a) (i) & (ii)

   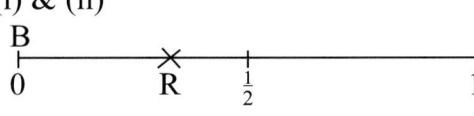

   (b) RG   RB   RP

       WG   WB   WP

       YG   YB   YP

3. (a) 6H 5H 4H 3H 2H 1H

       6T 5T 4T 3T 2T 1T

   (b) 1/12

4. (i) 1/5

   (ii)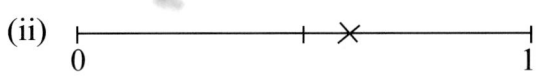

5. $\frac{25}{60} = \frac{5}{12}$

6. $\frac{31}{50}$

7. (a) $\frac{30}{100}$

   (b) 175

8. (a) 0.25

   (b) 60

Licence Agreement: *Edexcel Topic Tutor GCSE Maths Foundation, Student CD-ROM* (ISBN: 978-1-84690-007-5)

Warning:

This is a legally binding agreement between You (the user) and Edexcel Limited, of One90 High Holborn, London, WC1V 7BH, United Kingdom ("Edexcel").

By retaining this Licence, any software media or accompanying written materials or carrying out any of the permitted activities You are agreeing to be bound by the terms and conditions of this Licence. If You do not agree to the terms and conditions of this Licence, do not continue to use the *Edexcel Topic Tutor Student CD-ROM* and promptly return the entire publication (this Licence and all software, written materials, packaging and any other component received with it) with Your sales receipt to Your supplier for a full refund.

Edexcel Topic Tutor Student CD-ROM consists of copyright software and data. The copyright is owned by Edexcel. You only own the disk on which the software is supplied. If You do not continue to do only what You are allowed to do as contained in this Licence you will be in breach of the Licence and Edexcel shall have the right to terminate this Licence by written notice and take action to recover from you any damages suffered by Edexcel as a result of your breach.

Yes, You can:

1. use *Edexcel Topic Tutor Student CD-ROM* on your own personal computer as a single individual user.

No, You cannot:

1. copy *Edexcel Topic Tutor Student CD-ROM* (other than making one copy for back-up purposes);
2. alter the software included on the Edexcel *Topic Tutor Student CD-ROM*, or in any way reverse engineer, decompile or create a derivative product from the contents of the database or any software included in it (except as permitted by law);
3. include any software data from *Edexcel Topic Tutor Student CD-ROM* in any other product or software materials;
4. rent, hire, lend or sell *Edexcel Topic Tutor Student CD-ROM* to any third party;
5. copy any part of the documentation except where specifically indicated otherwise;
6. use the software in any way not specified above without the prior written consent of Edexcel.

Grant of Licence:

Edexcel grants You, provided You only do what is allowed under the Yes, You can section above, and do nothing under the No, You cannot section above, a non-exclusive, non-transferable Licence to use *Edexcel Topic Tutor Student CD-ROM*.

The above terms and conditions of this Licence become operative when using *Edexcel Topic Tutor Student CD-ROM*.

Limited Warranty:

Edexcel warrants that the disk or CD-ROM on which the software is supplied is free from defects in material and workmanship in normal use for ninety (90) days from the date You receive it. This warranty is limited to You and is not transferable.

This limited warranty is void if any damage has resulted from accident, abuse, misapplication, service or modification by someone other than Edexcel. In no event shall Edexcel be liable for any damages whatsoever arising out of installation of the software, even if advised of the possibility of such damages. Edexcel will not be liable for any loss or damage of any nature suffered by any party as a result of reliance upon or reproduction of any errors in the content of the publication.

Edexcel does not warrant that the functions of the software meet Your requirements or that the media is compatible with any computer system on which it is used or that the operation of the software will be unlimited or error free. You assume responsibility for selecting the software to achieve Your intended results and for the installation of, the use of and the results obtained from the software.

Edexcel shall not be liable for any loss or damage of any kind (except for personal injury or death caused by its negligence) arising from the use of *Edexcel Topic Tutor Student CD-ROM* or from errors, deficiencies or faults therein, whether such loss or damage is caused by negligence or otherwise.

The entire liability of Edexcel and your only remedy shall be replacement free of charge of the components that do not meet this warranty.

No information or advice (oral, written or otherwise) given by Edexcel's employees or agents shall create a warranty or in any way increase the scope of this warranty.

To the extent the law permits, Edexcel disclaims all other warranties, either express or implied, including by way of example and not limitation, warranties of quality and fitness for a particular purpose in respect of *Edexcel Topic Tutor Student CD-ROM*.

Governing Law:
This Licence will be governed and construed in accordance with English law.

© Edexcel Limited 2006